Evaluation of Cytotoxicity and Cytoprotection. Effects of Natural Toxins

Evaluation of Cytotoxicity and Cytoprotection. Effects of Natural Toxins

Editor

Ana Juan-García

MDPI • Basel • Beijing • Wuhan • Barcelona • Belgrade • Manchester • Tokyo • Cluj • Tianjin

Editor
Ana Juan-García
University of Valencia
Valencia
Spain

Editorial Office
MDPI
St. Alban-Anlage 66
4052 Basel, Switzerland

This is a reprint of articles from the Special Issue published online in the open access journal *Toxins* (ISSN 2072-6651) (available at: https://www.mdpi.com/journal/toxins/special_issues/cytotoxicity_cytoprotection).

For citation purposes, cite each article independently as indicated on the article page online and as indicated below:

LastName, A.A.; LastName, B.B.; LastName, C.C. Article Title. *Journal Name* **Year**, *Volume Number*, Page Range.

ISBN 978-3-0365-7090-7 (Hbk)
ISBN 978-3-0365-7091-4 (PDF)

Contents

toxins

MDPI

Editorial

Introduction to the *Toxins'* Special Issue on Evaluation of Cytotoxicity and Cytoprotection Effects of Natural Toxins

Ana Juan-García

Laboratory of Food Chemistry and Toxicology, Faculty of Pharmacy, University of Valencia, E-46100 Valencia, Spain; ana.juan@uv.es

The lifestyle associated with good nutritional quality of food is well known for its widely recognized health benefits, especially when rich in bioactive compounds. The classification of bioactive compounds is very wide, including lycopene, carotenoids, and polyphenols (flavonoids and non-flavonoids). Reduced risks of some types of cancer and other diseases have been associated with the adoption of such a diet, as well as having increased antioxidants, inhibitors of lipid peroxidation, a decrease in pro-inflammatory cytokine production, etc.

The presence of natural toxins in food usually happens due to a lack of good harvesting, storage, or packaging practices; climate changes; or atmospheric conditions. Such toxins can have different origins, such as from plants, fungi, algae, bacteria, marine biotoxins including mycotoxins, lectins, furocoumarins, Shiga toxin, ciguatoxins, etc.

The study of mycotoxins has increased in recent years and constitutes a great group to follow in the toxicology field due to their wide and dangerous effects in organs and systems and their presence in many foods, feed, and commodities. Mycotoxins are toxic secondary metabolites produced by filamentous fungi from *Fusarium*, *Alternaria*, and *Penicillium* spp. which spread naturally worldwide. Mycotoxins are also natural contaminants present in food and feed, and several health problems have been evidenced for both humans and animals. All in vitro and in vivo studies are key steps for risk assessment and the following legislation for mycotoxins.

An evaluation of the effects of natural toxins and biologically active compounds of extracts from the plant kingdom constitutes a potential to combat various diseases thanks to its rich content. The focus of this Special Issue of *Toxins* was to gather advances related to the cytotoxicity of natural toxins and the potential for the cytoprotection of natural compounds present in food or plants. In this context, this Special Issue of *Toxins* comprises eight original contributions.

Studies of extracts are presented in this issue from *Artemisia annua*, *Fridericia chica*, *Polygonum cuspidatum*, and *Coffea Arabica*, and its protective effects against mycotoxins, such as zearalenone (ZEN), α-zearalenone (α-ZEL), β-zearalenone (β-ZEL), beauvericin (BEA), and patulin (PAT). The following cell lines are used: human neuroblastoma cell line (SH-SY5Y), human embryonic kidney cells (HEK 293T), chicken granulosa cells, bovine aortic ECs (BAECs), bovine mammary epithelial cells (MAC-t), and fibroblast cells (3T3).

The Special Issue starts with the study of Alvarez-Ortega et al. [1] where the protective effects are reported of the hydroethanolic extract (HEFc) from the *Fridericia chica (Bignoniaceae)* leaves grown in the Colombian Caribbean against the cytotoxicity of α-ZEL and β-ZEL on SH-SY5Y cells. A determination of components in extracts through UPLC-QTOF-MS/MS is presented. HEFc showed a significant increase in cell viability after exposure to α-ZEL (25 and 50 μM) and β-ZEL (6–100 μM), and the HEFs are proposed as a valuable source of compounds with antioxidant properties against mycotoxins effects [1].

The Special Issue continues with a study of an extract from *Polygonum cuspidatum* rich in the antioxidant polydatin (PD) [2]. It was assayed in MAC-T cells against ZEN by Fu et al. [2]. It was revealed that ZEA + PD effectively reduced cell oxidative damage

Citation: Juan-García, A. Introduction to the *Toxins'* Special Issue on Evaluation of Cytotoxicity and Cytoprotection Effects of Natural Toxins. *Toxins* **2022**, *14*, 114. https://doi.org/10.3390/toxins14020114

Received: 20 December 2021
Accepted: 31 January 2022
Published: 2 February 2022

compared with the ZEA alone. Furthermore, it was observed that after qPCR analysis on ER stress-related genes and apoptosis genes (*Bax and Bcl-2*), PD down-regulated the expression of these genes as compared with ZEN as well as the caspase-3 activity. It was concluded that PD reduces ZEN-induced apoptosis by inhibiting oxidative damage and ER stress [2].

One of the main beverages drunk all over the world is coffee which is characterized by having a big amount of polyphenols. In addition, the sustainable objective of many companies in upcycling the waste products obtained from coffee is of great concern. Juan-García et al. [3] evaluated coffee extracts from coffee silverskin and spent coffee in a neuroblastoma cell line (SH-SY5Y cells) against beauvericin (BEA) and α-zearalenol (α-ZEL)-induced cytotoxicity with different strategies of treatment (direct, simultaneous, and pre-treatment strategies). Results were very diverse and opposite for some strategies of treatment, concentrations and mycotoxins assayed; however, there is a forthcoming promising use of these unexploited residues in the near future against mycotoxins effects [3].

In an attempt to know the different anti-cancer effects of *Artemisia annua*, extracts using a pressurized cyclic solid-liquid (PCSL) method (phytocomplex extracts) were compared with conventional extraction methods [4]. Effects were tested in the following cancer cell lines from humans, murine, and canines: Balb/c 3T3 mouse cells transformed by simian virus 40 (SVT2), Balb/c 3T3 mouse embryonic fibroblasts (NIH/3T3), human cervical cancer cells (HeLa), and canine osteosarcoma cells (CRL2130). Extracts were not assayed with any natural toxic compounds but evidenced the strong capacity to induce apoptosis and highlighted the possibility of using these extracts as a therapeutical strategy [4].

Studies of mycotoxins in animal cell lines are here presented for two of these cells against ZEN. Two studies in different cell lines are presented: in chicken granulosa cells and bovine aortic ECs (BAECs) [5,6]. The study in chicken granulosa cells was carried out by Zhu et al. [5] and was focused on studying the effect on the function of apoptosis and autophagy by gene expression (Bax, Bcl-2, Cyt C, and Caspase-9 and -3 for apoptosis; and LC3-II and Beclin-1 for autophagy). This cemented an understanding on the signaling pathway for autophagy activated by ZEN [5]. In bovine aortic ECs cells [6], a wide signaling pathway for ZEN was studied in apoptosis through caspase-3 and PARP (poly ADP-ribose polymerase) (ERK1/2/p53/caspase-3), but was independent of ROS production and estrogen receptor activation associated with ZEN.

In the same direction as indicated above, PAT was studied in HEK293 cells in producing and clearing ROS by Liu et al. [7]. This was carried out by studying genes involved in the mitochondrial respiratory chain complex and the role of NAC. It was revealed the importance of PAT in associated health problems [7].

Lastly, in the last chapter of this Special Issue, a validated study for biosafety has been carried out to decrease mycotoxicosis by Huang et al. [8]. NIH/3T3 mouse fibroblasts with products obtained from natural clays and denominated nano-silicate platelets (NSP) were used. ROS, caspase activation, and necroptosis signals have been studied. In conclusion, the study reports that these natural compounds (NPS) might have limited application due to the results obtained [8].

Funding: This research received no external funding.

Acknowledgments: The Guest Editor of this Special Issue, Ana Juan-García, is grateful to the authors for their contributions and particularly to the referees for their invaluable work. Without their effort, this Special Issue would have not been possible. The valuable contributions, organization, and editorial support of the MDPI management team and staff are greatly appreciated.

Conflicts of Interest: The author declares no conflict of interest.

References

1. Alvarez-Ortega, N.; Caballero-Gallardo, K.; Taboada-Alquerque, M.; Franco, J.; Stashenko, E.E.; Juan, C.; Juan-García, A.; Olivero-Verbel, J. Protective Effects of the Hydroethanolic Extract of Fridericia chica on Undifferentiated Human Neuroblastoma Cells Exposed to α-Zearalenol (α-ZEL) and β-Zearalenol (β-ZEL). *Toxins* **2021**, *13*, 748. [CrossRef] [PubMed]
2. Fu, Y.; Jin, Y.; Shan, A.; Zhang, J.; Tang, H.; Shen, J.; Zhou, C.; Yu, H.; Fang, H.; Zhao, Y.; et al. Polydatin Protects Bovine Mammary Epithelial Cells against Zearalenone-Induced Apoptosis by Inhibiting Oxidative Responses and Endoplasmic Reticulum Stress. *Toxins* **2021**, *13*, 121. [CrossRef] [PubMed]
3. Juan-García, A.; Caprioli, G.; Sagratini, G.; Mañes, J.; Juan, C. Coffee Silverskin and Spent Coffee Suitable as Neuroprotectors against Cell Death by Beauvericin and α-Zearalenol: Evaluating Strategies of Treatment. *Toxins* **2021**, *13*, 132. [CrossRef] [PubMed]
4. Culurciello, R.; Bosso, A.; Di Fabio, G.; Zarrelli, A.; Arciello, A.; Carella, F.; Leonardi, L.; Pazzaglia, L.; De Vico, G.; Pizzo, E. Cytotoxicity of an Innovative Pressurised Cyclic Solid–Liquid (PCSL) Extract from Artemisia annua. *Toxins* **2021**, *13*, 886. [CrossRef]
5. Zhu, Y.; Wang, H.; Wang, J.; Han, S.; Zhang, Y.; Ma, M.; Zhu, Q.; Zhang, K.; Yin, H. Zearalenone Induces Apoptosis and Cytoprotective Autophagy in Chicken Granulosa Cells by PI3K-AKT-mTOR and MAPK Signaling Pathways. *Toxins* **2021**, *13*, 199. [CrossRef]
6. Lee, H.-J.; Oh, S.-Y.; Jo, I. Zearalenone Induces Endothelial Cell Apoptosis through Activation of a Cytosolic Ca^{2+}/ERK1/2/p53/Caspase 3 Signaling Pathway. *Toxins* **2021**, *13*, 187. [CrossRef]
7. Liu, J.; Liu, Q.; Han, J.; Feng, J.; Guo, T.; Li, Z.; Min, F.; Jin, R.; Peng, X. N-Acetylcysteine Inhibits Patulin-Induced Apoptosis by Affecting ROS-Mediated Oxidative Damage Pathway. *Toxins* **2021**, *13*, 595. [CrossRef]
8. Huang, J.-T.; Chang, L.-C.; Cheng, C.-S.; Lin, J.-J.; Huang, S.-Y.; Chen, S.-E. Cytotoxicity Produced by Silicate Nanoplatelets: Study of Cell Death Mechanisms. *Toxins* **2020**, *12*, 623. [CrossRef] [PubMed]

toxins

MDPI

Article

Cytotoxicity Produced by Silicate Nanoplatelets: Study of Cell Death Mechanisms

Jie-Ting Huang [1,†], Ling-Chu Chang [2,3,4,†], Chung-Ssu Cheng [1], Jiang-Jen Lin [5], San-Yuan Huang [1,6,7,8] and Shuen-Ei Chen [1,6,7,8,*]

1 Department of Animal Science, National Chung Hsing University, Taichung 40227, Taiwan; jieting2283@gmail.com (J.-T.H.); terry80149@hotmail.com (C.-S.C.); syhuang@dragon.nchu.edu.tw (S.-Y.H.)
2 Chinese Medicinal Research and Development Center, China Medical University Hospital, Taichung 40447, Taiwan; t27602@mail.cmuh.org.tw
3 Center for Molecular Medicine, China Medical University Hospital, Taichung 40447, Taiwan
4 Department of Biological Science and Technology, China Medical University, Taichung 40402, Taiwan
5 Department of Materials Science and Engineering, National Chung Hsing University, Taichung 40227, Taiwan; jianglin@ntu.edu.tw
6 The iEGG and Animal Biotechnology Center, National Chung Hsing University, Taichung 40227, Taiwan
7 Innovation and Development Center of Sustainable Agriculture (IDCSA), National Chung Hsing University, Taichung 40227, Taiwan
8 Research Center for Sustainable Energy and Nanotechnology, National Chung Hsing University, Taichung 40227, Taiwan
* Correspondence: shueneic@dragon.nchu.edu.tw; Tel.: +886-4-2287-0613
† These two authors contribute equally to this work.

Received: 26 August 2020; Accepted: 29 September 2020; Published: 29 September 2020

Abstract: Nano-silicate platelets (NSP), an exfoliated product from natural clays, have been validated for biosafety and as an effective supplement to alleviate mycotoxicosis. Since NSP induced noticeable cell death, we therefore investigated further the mechanism of cytotoxicity caused by NSP. Exposure to NSP impaired membrane integrity and caused cell death in a dose-dependent manner. Reactive oxygen species (ROS) generation other than of NADH oxidase origin, and subcellular interactions by internalized NSP also contributed to NSP-induced cell death. NSP persistently provoked receptor-interacting protein 1 Ser/Thr (RIP1) kinase and caspase 6 and 3/7 activation without altering caspase 8 activity and induced evident chromatolysis of necrosis in the later stage. These events proceeded along with increased ER stress and mitochondrial permeability, to final Cyt-C (Cytochrome C) release and AIF (apoptosis inducing factor) translocation, a hallmark of cell necroptosis. Fluorescent probing further manifested NSP traffic, mostly adherence on the cell surfaces, or via internalization, being compartmentalized in the nuclei, cytosols, and mitochondria. Pharmacological approaches with specific inhibitors suggested that endocytosis and particularly RIP1 kinase provocation mediate NSP-induced cell death independent of caspase activation. In conclusion, the necroptotic process contributes to most of the cell death induced by NSP due to membrane interactions/impaired integrity, ROS generation, and subcellular interactions by internalized NSP.

Keywords: nano-silicate platelets; necroptosis; reactive oxygen species; endocytosis; membrane integrity

Key Contribution: The necroptotic process accounts for most of the cell death induced by NSP.

1. Introduction

Conventionally, sheet silicate minerals from phyllosilicates, including bentonite and montmorillonite clays, have been used as medical materials for diarrhea in humans, due to their antibacterial activity [1]. In livestock management, these aluminosilicate clays, as well as purified sodium calcium aluminosilicates (HSCAS), have been widely used as a supplement to alleviate mycotoxicosis in poultry and swine [2].

Aluminosilicate-containing clays exhibit a strong interaction attracting polar functional groups of mycotoxins depending on pore size, charge intensity, and intrinsic affinity and thereby hinder the absorption of mycotoxins in the digestive tract [3,4]. The core aluminum ions of HSCAS can interact with the β-carbonyl group of aflatoxins (AFB1) leading to a high affinity to AFB1 [5]. Despite being quite effective in detoxifying aflatoxicosis, HSCAS products are less effective in counteracting toxicosis by *Fusarium* mycotoxins, such as fumonisins, trichothecenes, and zearalenone [2].

Previously, we demonstrated that exfoliated sheet-like aluminosilicate clays serve as an effective feed supplement to alleviate marker pathologies and improve growth performance in chickens intoxicated with fumonisin B1 (FB1) or AFB1 [6,7]. As low as 40–200 mg/kg feed of exfoliated silicate nanoplatelets were able to counteract toxicosis by FB1 and AFB1. Dietary inclusion of nano-silicate platelets (NSP) also promoted feed intake and growth in chickens as well as the alleviation of FB1 toxicosis [6]. In pregnant mice intoxicated with FB1, oral administration of NSP at 2.5 mg/kg BW suppressed maternal plasma FB1 concentrations by 93%, ameliorated embryo neural tube defects and teratogenesis, and thus significantly improved fetus growth [8].

The exfoliative process with aluminosilicate clays including montmorillonite and bentonite afforded NSP nano-clays that are finely dispersible in water and possess physical properties of a polygonal geometric shape of one nanometer thickness (ca. 80 × 80 × 1 dimension) with high surface area (ca. 720 m^2/g), cationic exchange capacity at 1.20 meq/g, and intensive ionic charges (ca. 20,000 ions/platelet) [9–11]. These unique characteristics endow NSP with a high affinity to adhere to the surface of bacteria, and thereby with a strong bacteriostatic and bactericidal activity [11–15]. Activation of death signaling and reactive oxygen species (ROS) provocation irrelevant to Ag^+ cytotoxicity was further shown to mediate the bactericidal effects of AgNP (silver nanoparticle)/NSP nanohybrids on silver-resistant *E. coli* [13,14,16]. Surfactants were also used to modify NSP at noncytotoxic concentrations and provoke a broad and potent antiviral activity due to electrostatic interactions with the virus in order to block the viral access to cell surfaces [17]. In in vivo studies, sodium dodecyl sulfate (SDS) surfactant facilitated NSP in shielding off infection by dengue, Japanese encephalitis, and influenza A virus, resulting in a reduction of lethality in the infected mice [17].

For practical applications, NSP was evaluated for biosafety and showed a very low toxicity, the lethal dose (LD50) being >5700 mg/kg of body weight in an acute toxicity study, in which rats receiving daily oral administration of NSP for 2 weeks exhibited normal livability, body weight change, feed intake, behaviors, and histology [12]. Despite the absence of genotoxicity, a very low but noticeable induction of cell death was observed when cells were exposed to NSP at levels > 62.5 μg/mL [12]. For further understanding of the safety issues, we embarked on a thorough investigation of the cytotoxicity and the mechanistic aspects of cell death by NSP induction.

2. Results

2.1. Cell Death by NSP Exposure

Exposure to NSP induced cell death in a dose-dependent manner, in which early apoptosis accounted for most cell death (10.1 and 13.1% at 24 and 48 h by 200 μg/mL, respectively), and interestingly only late apoptosis increased during the time course (3.7 to 7.8%), whereas necrosis contributed to cell death less than 2% ($p < 0.05$, Figure 1, panel A). NSP treatment for 24 h also increased cellular LDH (lactate dehydrogenase) leakage in a dose-dependent manner, suggesting damaged membrane integrity ($p < 0.05$, Figure 1, panel B). Instead of typical laddering fragmentation of DNA breakdown in apoptosis, cell death proceeded to necrotic chromatinolysis with a more pronounced smearing pattern of DNA size in electrophoresis after exposure to NSP for 48 h (Figure 1, panel C).

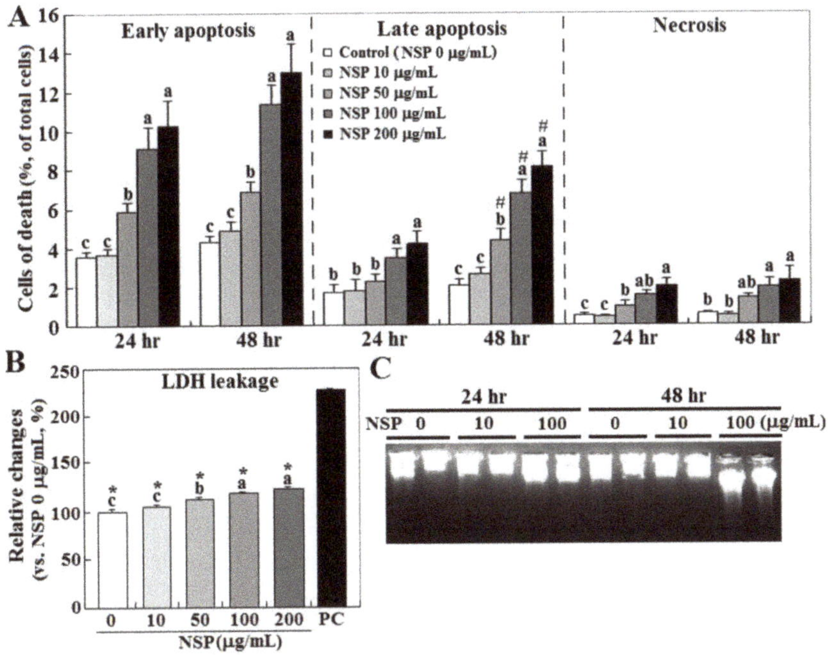

Figure 1. Effects of nano-silicate platelets (NSP) on cell death, lactate dehydrogenase (LDH) leakage, and DNA breakdown. Cells incubated with various levels of NSP for 24 or 48 h were collected for cell death (panel **A**), LDH leakage (24 h, panel **B**), and DNA breakdown analysis (panel **C**) (n = 4). Means with different superscript letters differ significantly among various levels of NSP treatment ($p < 0.05$). #; significant difference vs. 24 h treatment within the same type of cell death ($p < 0.05$). *; significant difference vs. positive control in LDH leakage ($p < 0.05$). Medium and cell lysates from the cultures without NSP supplementation (0 μg/mL) serve as the negative and positive (PC) control, respectively.

2.2. *ROS Contribution and Origins*

Treatment of PTDC (ammonium pyrrolidinedithiocarbamate) and n-MPG (n2-mercaptopropionyl-glycine) for ROS scavenging, or Cyto D (cytochalasin D) to block endocytosis, but not by NADH (nicotinamide adenine dinucleotide) oxidase inhibitors, DPI (diphenyleneiodonium chloride) or Apo (apocyni), partially rescued cell death induced by NSP ($p < 0.05$, Figure 2, panel A). Exposure to NSP induced ROS production regardless of the presence of the pharmacological inhibitors ($p < 0.05$, Figure 2, panel B). In contrast to their respective vehicle control, ROS production was suppressed by DPI but increased by Apo in both vehicle control and NSP-treated cells ($p < 0.05$, Figure 2, panel B), consistently with previous studies which showed that Apo represses NADH oxidase activity only in phagocytic cells, but stimulates ROS generation in non-phagocytic cells [18]. In contrast to the vehicle control, Cyto D treatment suppressed ROS generation in NSP-free conditions ($p < 0.05$, Figure 2, panel B), but not in the presence of NSP. In combination with results from Figure 1, ROS generation other than by NADH oxidase origin, and intracellular interactions/mechanisms activated by internalized NSP irrelevant to ROS generation, were concluded to contribute to NSP-induced cell death.

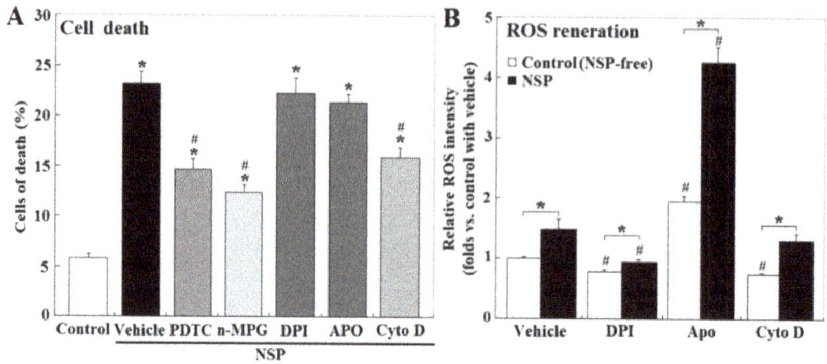

Figure 2. Effects of reactive oxygen species (ROS) scavenging and blockade of NADH oxidase and endocytosis on NSP-induced cell death and ROS production. Cells pre-treated with vehicle or pharmacological inhibitors were incubated with 100 μg/mL NSP for 48 h and then harvested for cell death analysis (panel **A**) or for 3 h for ROS generation analysis (panel **B**) (n = 4). Results of cell death were combined with early, late apoptosis and necrosis. *; significant difference by NSP treatment (vs. control, $p < 0.05$). #; significant difference by inhibitor treatment (vs. corresponding group in vehicle treatment, $p < 0.05$). PTDC; ammonium pyrrolidinedithiocarbamate, n-MPG; n2-mercaptopropionyl-glycine, Cyto D; cytochalasin D, DPI; diphenyleneiodonium chloride, Apo; apocynin, NADH; Nicotinamide adenine dinucleotide.

2.3. RIP1 Kinase, Caspase Activation, and ER Stress

NSP exhibited a high affinity to adhere onto cell surfaces [12] and induced cell death with necrotic chromatinolysis rather than laddering fragmentation of DNA breakdown in apoptosis (Figure 1). Membrane integrity impairment/interactions, ROS provocation, and the endocytic process were shown to contribute to NSP-induced cell death (Figures 1 and 2). We then studied the type of cell death and its progression.

Exposure to NSP persistently promoted RIP1 (receptor-interacting protein 1 Ser/Thr) kinase activation and caspase 6 and 3/7 activity, two executioners in cell apoptosis, but not in caspase 8 ($p < 0.05$, Figure 3, panel A and B). The induction of caspase 6 and 3/7 activity by NSP was completely reversed by Cyto D but not by n-MPG, whereas neither n-MPG nor Cyto D affected caspase 8 activity ($p < 0.05$, Figure 3, panel C and D). Since caspase 8 mediates the extrinsic apoptotic signaling pathway and acts as a RIP1 kinase repressor, whose activity determines cell death by apoptotic or necroptotic process [19,20], these results thus exclude the activation of the extrinsic apoptotic pathway and suggest the necroptotic process and intrinsic apoptosis involved in the progression of NSP-induced cell death. It is concluded that ROS generation mediates NSP-induced cell death in a caspase-independent manner. Treatment of n-MPG and Nec-1 (Necrostatin-1, a RIP1kinase inhibitor) relieved ER (endoplasmic reticulum) stress by NSP as evidenced by downregulation of BiP and CHOP (C/EBP homologous protein) expression ($p < 0.05$, Figure 4). Blockade of endocytosis by Cyto D completely abolished the increases of ER stress caused by NSP ($p < 0.05$, Figure 4), suggesting that physical interactions with subcellular organelles or components such as proteins, and subsequent mechanisms provoked by internalized NSP, promote ER stress.

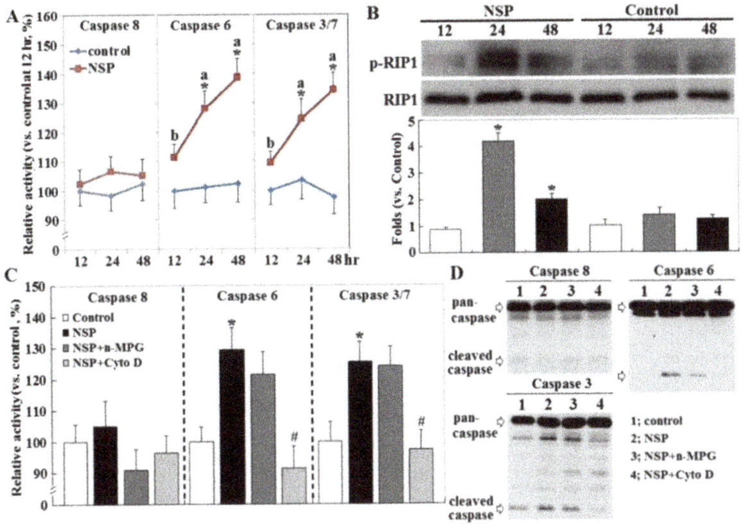

Figure 3. Receptor-interacting protein 1 Ser/Thr (RIP1) activation, and effect of ROS scavenging and endocytosis blockade on NSP-induced caspase activation. Cells alone (panel **A**), or pre-treated with vehicle or pharmacological inhibitors, were incubated with 100 μg/mL NSP for 12, 24 or 48 h and then harvested for caspase (panel **C** at 24 h, panel **D** at 48 h) and RIP activation (panel **B**) analysis through enzymatic or Western blot method (n = 4). Means with different superscript letters differ significantly among different time points ($p < 0.05$). *; significant difference by NSP treatment (vs. corresponding control, $p < 0.05$). #; significant difference by inhibitor treatment (vs. NSP treatment, $p < 0.05$). n-MPG; n2-mercaptopropionyl-glycine, Cyto D; cytochalasin.

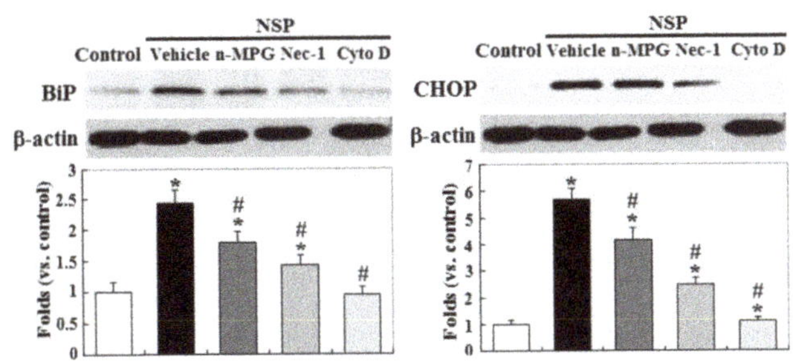

Figure 4. Effects of ROS scavenging and blockade of endocytosis and RIP1 kinase on SP-induced ER stress. Cells pre-treated with vehicle or pharmacological inhibitors were incubated with 100 μg/mL NSP for 24 h and then harvested for BiP and CHOP (C/EBP homologous protein) expression through Western blot analysis (n = 4). *; significant difference by NSP treatment (vs. control, $p < 0.05$). #; significant difference by inhibitor treatment (vs. vehicle with NSP treatment, $p < 0.05$). n-MPG; n2-mercaptopropionyl-glycine, Cyto D; cytochalasin D, Nec-1; necrostatin-1.

2.4. Mitochondrial Membrane Potential, Cyt-C Release, and AIF Translocation

Treatment with Nec-1 and Cyto D, but not by n-MPG, differentially rescued downregulation of gelsolin by NSP exposure, a mitochondrial permeability stabilizer ($p < 0.05$, Figure 5) [21]. Downregulation of HSP70 (heat shock protein 70) by NSP, a protein chaperone which repressively interacts with caspase 9 and cytosolic AIF (apoptosis inducing factor) [22], was completely reversed by Cyto D and ameliorated by Nec-1 and n-MPG ($p < 0.05$). Consistent with these results, n-MPG, Cyto D, and Nec-1 differentially ameliorated mitochondrial membrane potential (MMP) loss by NSP, i.e., by rescuing mitochondrial membrane permeability ($p < 0.05$, Figure 5) and attenuating the downstream events including mitochondrial Cyt-C release, a critical activator for caspase cascading in intrinsic apoptosis, and AIF translocation into the nuclei where it induces chromatolysis, a hallmark of cell necroptosis ($p < 0.05$, Figure 6) [23].

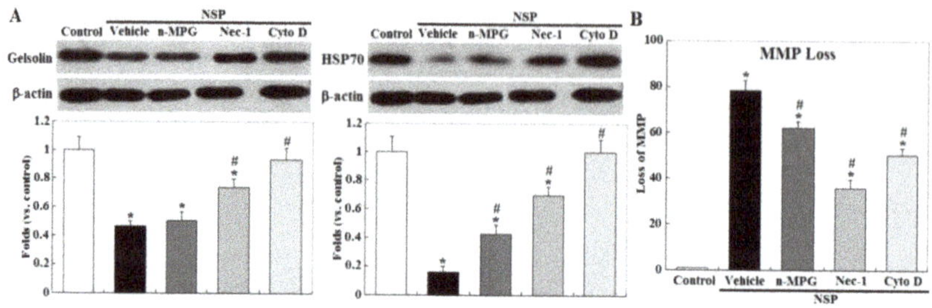

Figure 5. Effects of ROS scavenging and blockade of endocytosis and RIP1 kinase on NSP-induced expression of gelsolin, heat shock protein 70 (HSP70) and mitochondrion membrane potential. Cells pre-treated with vehicle or pharmacological inhibitors were incubated with 100 µg/mL NSP for 24 h and then harvested for gelsolin and HSP70 expression by Western blot (panel **A**) or for MMP (mitochondrial membrane potential) (panel **B**) analysis (n = 4). *; significant difference by NSP treatment (vs. control, $p < 0.05$). #; significant difference by inhibitor treatment (vs. vehicle with NSP treatment, $p < 0.05$). n-MPG; n2-mercaptopropionyl-glycine, Cyto D; cytochalasin D, Nec-1; necrostatin-1.

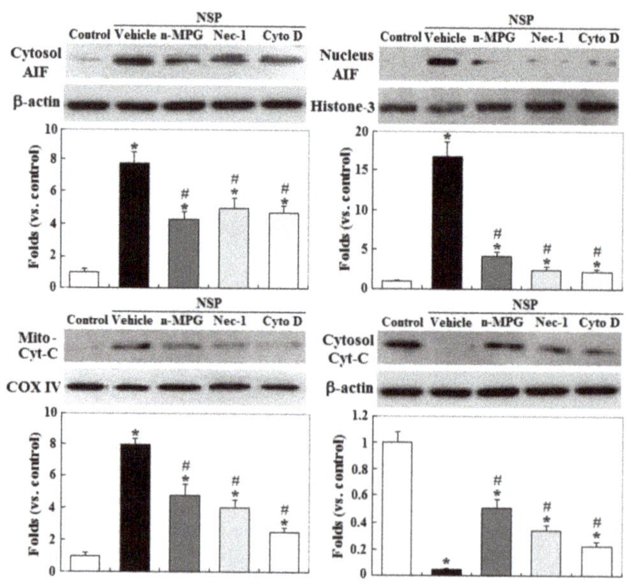

Figure 6. Effects of ROS scavenging and blockade of endocytosis and RIP1 kinase on NSP-induced translocation of AIF (apoptosis inducing factor) and Cyt-C (cytochrome C) release. Cells pre-treated with vehicle or pharmacological inhibitors were incubated with 100 µg/mL NSP for 48 h and then harvested for fractionation for AIF and Cyt-C translocation analysis through Western blot method (n = 4). *; significant difference by NSP treatment (vs. control, $p < 0.05$). #; significant difference by inhibitor treatment (vs. vehicle with NSP treatment, $p < 0.05$). n-MPG; n2-mercaptopropionyl-glycine, Cyto D; cytochalasin D, Nec-1; necrostatin-1.

2.5. Localization of NSP Traffic

Morphological probing with fluorescent NSP-APTES-FITC localized NSP mainly aggregated to adherence onto the cell surfaces (Figure 7, panel A). Some NSP were ingested and trafficked to associate with the nuclei and mitochondria or scattered in the cytosols (panel B to D).

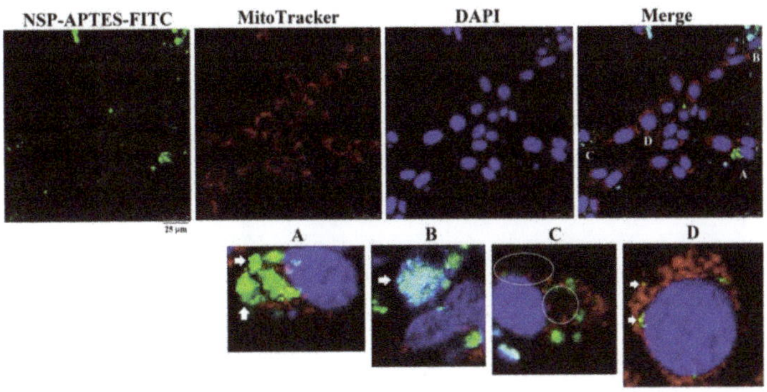

Figure 7. Subcellular compartmentalization of NSP. Cells were incubated with NSP-APTES-FITC (10 μg/mL) for 24 h. MitoTracker® Red CMXRos was used to probe mitochondria. NSP-APTES-FITC were localized on the cell surfaces (panel **A**, bright green color), nuclei (panel **B**, cyan color), cytosol (panel **C**, scattered faint green color), and mitochondria (panel **D**, yellow color) as indicated by arrows or within circles. APTES; (3-aminopropyl) triethoxysilane, FITC; fluorescein isothiocyanate.

2.6. Caspase and RIP1/3 Kinase Activation, and NSP Internalization in NSP-Induced Cell Death

Surprisingly, zVAD-fmk (Z-Val-Ala-Asp (OMe)-fluoromethyl ketone, a broad spectrum caspase inhibitor) failed, but Nec-1 partially rescued, NSP-induced cell death ($p < 0.05$, Figure 8). zVAD-fmk+Cyto D and zVAD-fmk+Nec-1 treatment also ameliorated NSP-induced cell death to a higher level than zVAD-fmk or Nec-1 treatment alone ($p < 0.05$), suggesting that necroptosis through RIP1/3 activation accounts for most NSP-induced cell death, whereas caspase activation and particularly internalized NSP synergistically participate in the programmed cell death. Treatment with Nec-1+Cyto D completely abolished cell death by NSP ($p < 0.05$).

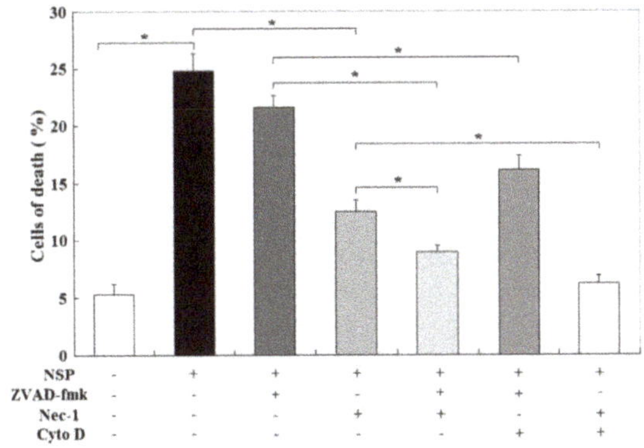

Figure 8. Effects of caspase, endocytosis, and RIP kinase blockade on NSP-induced cell death. Cells pre-treated with vehicle or pharmacological inhibitors were incubated with 100 μg/mL NSP for 48 h and then harvested for canalysis. Results of cell death were combined with early, late apoptosis and necrosis. *; significant difference ($p < 0.05$). Cyto D; cytochalasin D, Nec-1; necrostatin-1, ZVAD-fmk; Z-Val-Ala-Asp (OMe)-fluoromethyl ketone. +, −; present or absent with the indicated chemical(s).

3. Discussion

The present study defined cytotoxicity by NSP exposure including ROS induction, cell membrane interaction/damage, and intracellular interactions/mechanisms caused by internalized NSP in mediating cell death. This process of cell death evolved with RIP1 kinase activation in a caspase-independent manner, proceeded by downregulation of HSP70 and gelsolin, increased ER stress and mitochondrial permeability, and finally activated AIF translocation leading to cell necroptosis.

Most cell death by NSP exhibited apoptotic hallmarks including PS (phosphatidyl serine) exposure (annexin V-positive/Propidium iodide I-negative), mitochondrial Cyt-C release and caspase 6 and 3/7 activation. However, cells underwent death in the presence of zVAD-fmk and were sensitive to Nec-1, consistent with AIF translocation into the nuclei, necrotic smearing of DNA breakdown instead of laddering fragmentation, and no changes in caspase 8 activity.

Upon the activation of death receptors, caspase 8 repressively binds to RIP1 kinase and thus allows activation of apoptotic caspase signaling, but when caspase 8 is inhibited or activated inefficiently, RIP1 kinase can interact with RIP3 and drive cells into necroptosis [19]. Accordingly, these results suggest that the apoptotic pathway only accounts for a small part of NSP-induced cell death, which mostly occurs through necroptosis, a type of regulated cell death sharing some apoptotic phenotypes, but characterized by RIP1/3 kinase activation, AIF translocation into nuclei, and necrotic chromatolysis [19,20]. Apoptosis-like cell death, insensitive to zVAD-fmk but exacerbated by specific inhibition of caspase-8 activity, has been observed in NIH/3T3 cells in response to death receptor signaling [22] and in HepG2 cells induced by unmodified silicate clays [24].

Consistent with previous reports [11–14], a large portion of NSP was observed to adhere onto the cell membranes. This adherence may alter osmolality and polarization in local regions leading to impaired membrane integrity, provoke ROS generation, interact with local molecules such as death receptors, and thus activate death signaling [25]. Hyperpolarization of plasma membranes was shown to open calcium channels and resultant calcium cascades leading to ROS generation, calpain, and JNK (c-Jun N-terminal kinases) activation, which further act on mitochondrial PTPC (permeability transition pore complex) leading to AIF release into the cytosol, where AIF is activated by calpain processing [23,26].

In addition to functioning as a chaperone for proper protein folding and misfolded protein removal to relieve ER stress [27], HSP70 specifically represses AIF translocation and sequesters procaspase-9 from activation [23]. Gelsolin acts as a mitochondrial permeability stabilizer to prevent Cyt-C release [21]. Accordingly, downregulation of HSP70 and gelsolin by NSP may promote ER stress and facilitate AIF translocation and Cyt-C release to steer cells into the necroptotic process [28]. ROS provocation, other than by NADPH oxidase, mediates parts of NSP-induced cell death regardless of caspase activation, but is operative in mitochondrial permeability and AIF release. These results can be attributed to activation of necroptotic pathway, in which RIP1/3 kinase interacts with several mitochondrial metabolic enzymes and enhances autophagic degradation of catalase leading to ROS overproduction, and thereby impaired mitochondrial permeability [26].

Endocytosis blockade by Cyto D has been shown to ameliorate cytotoxicity involving ROS generation by amorphous nano-silica particles [29]. Non-phagocytic cells tend to ingest cationic nanoparticles depending on charge density and hydrophobicity [30]. As evidenced in the morphological probing study, high surface ionic charges of NSP promoted NSP ingestion into cells, and thus increased physical interactions with intracellular organelles leading to ROS generation and mitochondrial disruption, and subsequently cell death cascading. Physical interactions may also directly damage protein and DNA structure and translation machinery leading to ER stress, proteotoxicity and genotoxicity, when viewing the literature reporting the modification of nano-clays for susceptibility of ingestion by cells [30]. Ingestion of organically modified nano-clays has been shown to present in cytoplasmic vesicles and participate in cell death [31].

4. Conclusions

The present study concluded that ROS provocation, cell membrane interaction/damage, and intracellular interactions/mechanisms by internalized NSP mediate NSP-induced cell death. These cytotoxic processes operate by RIP1 kinase activation, ER stress and mitochondrial permeability, leading to AIF activation and ultimately cell necroptosis.

5. Materials and Methods

5.1. Preparations of NSP

Nanosilicate platelets were prepared from natural sodium montmorillonite (Na^+-MMT) by exfoliation of the layered silicate clays using home-made polyamine-HCl as the exfoliating agent. Details of the synthetic procedures, purification, and characteristics of the NSP products were described previously [9–11].

5.2. Cell Cultures

NIH/3T3 mouse fibroblasts (ATCC, Manassas, VA, USA) were cultured in DMEM medium (Gibco, New York, NY, USA) supplemented with 10% fetal calf serum (Gibco) and 1% penicillin–streptomycin solution (Gibco), pH 7.4 at 37 °C, 5% CO2, 95% humidity. Medium was changed every 2 days. When reaching 80–90% confluence, cells were pre-treated with various pharmacological inhibitors for different durations including n-MPG (in distilled water, final concentrations 300 μM, Sigma-Aldrich, St. Louis, MN, USA), PDTC (in distilled water, 2 mM, Sigma-Aldrich), DPI (in DMSO, 10 μM, Calbiochem, San Diego, CA, USA), Apo (in DMSO, 30 μM, Calbiochem), and Cyto D (in DMSO, 2 nM, Sigma-Aldrich) for 30 min, zVAD-fmk (in DMSO, 50 μM, BioVision, Milpitas, CA, USA) for 1 h, and Nec-1 (in DMSO, 10 μM, BioVision) for 2 h. After replacing with new medium, cells were cultured with NSP and collected at indicated time points for further analyses. NSP were dispersed in PBS buffer with sonication for 10 min prior to treatment.

5.3. Cell Death and LDH Leakage Analysis

Cell death was analyzed by annexin-V/PI (propidium iodide) method using a commercial kit (BD Biosciences, San Jose, CA, USA) and flow cytometry for cell death sorting (FC500, Beckman Coulter Inc. Brea, CA, USA) within 1 h after staining [32,33]. Genomic DNA extracts were used for chromatinolysis analysis by evaluating DNA degradation under electrophoresis in 2% agarose gels. Cell lysates and culture medium were collected for LDH activity analysis using a commercial kit (Promega, Madison, WI, USA). Cell lysates and collected medium without NSP treatment were used as positive and negative control, respectively.

5.4. ROS Production, Mitochondrial Membrane Potential, and Caspase Activity

Cells pre-treated with various inhibitors or vehicles were cultured with 25 μM 2′,7′-dichlorofluorescin diacetate (DCFDA, Abcam, Cambridge, UK) at 37 °C in dark for 45 min. The cultures were then replaced with phenol red-free medium containing 100 μg/mL NSP and incubated for another 3 h. Generation of ROS was determined by fluorescence intensity with Ex/Em = 485/535 nm.

Determination of mitochondrial membrane potential (MMP, $\Delta\psi m$) was conducted as described previously [34] using 3,3′-dihexyloxacarbocyanine iodide [DiOC6(3), Molecular Probes Inc, Eugene, OR, USA] as a molecular dye. In brief, cells pre-treated with various inhibitors or vehicles were incubated with 100 μg/mL NSP for 48 h. Collected cells were suspended in PBS buffer containing DiOC6(3) (first dissolved in DMSO and diluted in PBS buffer, final concentration 40 nM). After incubation at 37 °C in dark for 30 min, cells were pelleted and washed with PBS to remove DiOC6(3). After 3 washes, MMP was measured by flow cytometric analysis (FC500, Beckman Coulter Inc.).

Caspase 6, 3/7, and 8 activities were determined as the cleavage rate of the synthetic fluorophoric (Biovision, Promega) or chromophoric (Invitrogen, Carlsbad, CA, USA) peptide substrate, respectively. Caspase 6 and 3/7 activity were read in with a microplate reader (Infinite F200 PRO, Tecan Group Ltd., Mannedorf, Switzerland) at Ex/Em = 400/505 and 480/520 nm, respectively, and caspase 8 activity was read at 400 nm.

5.5. Western Blot Analysis

Total cell lysates were prepared in RIPA (radioimmunoprecipitation assay) buffer containing protease and phosphatase inhibitor cocktail (Sigma-Aldrich). Isolation of cytoplasmic and nuclear fractions was conducted with the hypotonic buffer and centrifugation method as described previously [35]. Mitochondrial fractions were isolated according to the instructions enclosed in the commercial kit from Abacm. In electrophoresis for protein separation, each well contained a respective sample with 60 μg of proteins from total cell lysates, or 15–20 μg of proteins from cytoplasmic or nuclear fractions. Proteins were transferred onto PVDF (polyvinylidene fluoride) membrane with the wet-transfer method. A mouse anti-HSP70 (clone N27F3-4) monoclonal antibody was purchased from Enzo Life Sciences (New York, NY, USA). Rabbit anti-gelsolin (clone D9W8Y), anti-Cyt-C (clone 136F3), anti-COX IV (cytochrome c oxidase IV, clone 3E11), anti-caspase 8 (clone 1C12), anti-caspase 3 (clone D3R6Y), anti-RIP1 (clone D94C12) anti-phospho-RIP1 (clone D1L3S), anti-BiP (cone C50B12), anti-CHOP (clone D46F1), and anti-anti-β-actin (clone 13E5) monoclonal antibody were purchased from Cell Signaling Technologies (Danvers, MA, USA). A rabbit polyclonal anti-AIF, anti-histone 3, and anti-caspase 6 were derived from Aviva Systems Biology (San Diego, CA, USA), Abcam, and Cell Signaling Technologies, respectively. Horseradish peroxidase-conjugated secondary antibodies; goat anti-mouse IgG (Beckman Coulter, Brea, CA, USA) and anti-rabbit IgG (Calbiochem) were used for to identify the bands reactive to the primary antibodies through an enhanced chemiluminescence reagent (Pierce Biotechnology Inc., Rockford, IL, USA). Primary and secondary antibodies were incubated with membranes at 1:1000 and 1:7500–10,000 dilation, respectively.

5.6. Subcellular Compartmentalization of NSP

To visualize NSP traffics, NSP was tethered with biocompatible APTES dependents through alkaline catalysis by NH4OH (pH 11) to form NSP-APTES. Then, FITC was attached onto NSP-PHEMA to form NSP-APTES-FITC. Details of the preparations of NSP-APTES-FITC were described previously [36].

Cells grown on coverslips in a 6-well plate were cultured with NSP-APTES-FITC (10 μg/mL) for 24 h. Prior to cell collection, cells were incubated with MitoTracker® Red CMXRos (Cell Signaling Technologies, Danvers, MA, USA) for 30 min and then fixed with 3.7% (*w/v*) paraformaldehyde and permeabilized with 0.2% (*v/v*) Triton X-100 for the determination of mitochondrial morphology. Coverslips were mounted and fluorescence images were taken by a Leica Microsystems TCS SP8 Confocal Spectral microscope (Leica Microsystems, Wetzlar, Germany) with Ex/Em = 579/599 nm for MitoTracker® Red CMXRos and 493/528 nm for FITC.

5.7. Statistics

Data were analyzed by one-way ANOVA, in which NSP or pharmacological inhibitor treatments, or time point, were the classifying variables as indicated. Result were expressed as means ± SE. Differences between means were tested using Student's *t*-test or Duncan's multiple comparison procedure. All statistical procedures were performed by using SAS software (2000). Statistical significance was accepted at $p < 0.05$.

Author Contributions: Conceptualization, S.-E.C., J.-J.L. and S.-Y.H.; methodology, S.-E.C., J.-J.L.; software, S.-E.C.; validation, S.-E.C. and J.-J.L.; formal analysis, S.-E.C., J.-J.L.; investigation, J.-T.H., L.-C.C., and C.-S.C.; resources, S.-E.C., J.-J.L. and L.-C.C.; data curation, S.-E.C., L.-C.C.; writing—original draft preparation, S.-E.C., J.-J.L. and S.-Y.H.; writing—review and editing, S.-E.C., and J.-J.L.; visualization, S.-E.C.; supervision, S.-E.C.; project administration, S.-E.C. and L.-C.C.; funding acquisition, S.-E.C., and L.-C.C. All authors have read and agreed to the published version of the manuscript.

Funding: This work was financially supported by the grant of the Ministry of Science and Technology, Taiwan; MOST 107-2313-B-005-036-MY3, the iEGG and Animal Biotechnology Center and the Innovation and Development Center of Sustainable Agriculture from The Featured Areas Research Center Program within the framework of the Higher Education Sprout Project by the Ministry of Education (MOE) in Taiwan to S.-E. Chen, and by the Ministry of Science and Technology, Taiwan; MOST 108-2320-B-039-040, MOST 109-2320-B-039-016, China Medical University, Taiwan; CMU108-MF-54, and China Medical University Hospital, Taiwan; DMR-109-131 to L.-C. Chang.

Acknowledgments: We appreciate Ru-Chun Tai, the Medical Research Core Facilities Center, Office of Research and Development at China Medical University, Taichung, Taiwan, for helping and assisting with the Confocal Spectral microscope.

Conflicts of Interest: The authors declare no conflict of interest.

References

1. Gelineau-van Waes, J.; Starr, L.; Maddox, J.; Aleman, F.; Voss, K.A.; Wilberding, J.; Riley, R.T. Maternal fumonisin exposure and risk for neural tube defects: Mechanisms in an in vivo mouse model. *Birth Defects Res. Part. Clin. Mol. Teratol.* **2005**, *73*, 487–497. [CrossRef]
2. Galvano, F.; Piva, A.; Ritieni, A.; Galvano, G.J. Dietary strategies to counteract the effects of mycotoxins: A review. *Food Prot.* **2001**, *64*, 120–131. [CrossRef]
3. Huwig, A.; Freimund, S.; Kappeli, O.; Dutler, H. Mycotoxin detoxification of animal feed by different adsorbents. *Toxicol. Lett.* **2001**, *122*, 179–188. [CrossRef]
4. Dixon, J.B.; Kannewischer, I.; Tenorio Arvide, M.G.; Barrientos Velazquez, A.L. Aflatoxin sequestration in animal feeds by quality-labeled smectite clays: An introductory plan. *Appl. Clay Sci.* **2008**, *40*, 201–208. [CrossRef]
5. Phillips, T.D. Dietary clay in the chemoprevention of aflatoxin induced disease. *Toxicol. Sci.* **1999**, *52*, 118–126. [CrossRef]
6. Yuan, C.W.; Huang, J.T.; Chen, C.C.; Tang, P.C.; Huang, J.W.; Lin, J.J.; Huang, S.Y.; Chen, S.E. Evaluation of Efficacy and Toxicity of Exfoliated Silicate Nanoclays as a Feed Additive for Fumonisin Detoxification. *J. Agric. Food Chem.* **2017**, *65*, 6564–6571. [CrossRef]
7. Tzou, Y.M.; Chan, Y.T.; Chen, S.E.; Wang, C.C.; Chiang, P.N.; Teah, H.Y.; Hung, J.T.; Wu, J.J.; Liu, Y.T. Use 3-D tomography to reveal structural modification of bentonite-enriched clay by nonionic surfactants: Application of organo-clay composites to detoxify aflatoxin B1 in chickens. *J. Hazard. Mater.* **2019**, *375*, 312–319. [CrossRef]
8. Liao, Y.J.; Yang, J.R.; Chen, S.E.; Wu, S.J.; Huang, S.Y.; Lin, J.J.; Chen, L.R.; Tang, P.C. Inhibition of fumonisin B1 cytotoxicity by nanosilicate platelets during mouse embryo development. *PLoS ONE* **2014**, *9*, e112290. [CrossRef]
9. Chu, C.C.; Chiang, M.L.; Tsai, C.M.; Lin, J.J. Exfoliation of montmorillonite clay by mannich polyamines with multiple quaternary salts. *Macromolecules* **2005**, *38*, 6240–6243. [CrossRef]
10. Chiu, C.W.; Huang, T.K.; Wang, Y.C.; Alamani, B.G.; Lin, J.J. Intercalation strategies in clay/polymer hybrids. *Prog. Polym. Sci.* **2014**, *39*, 443–485. [CrossRef]
11. Hsu, S.H.; Tseng, H.J.; Hung, H.S.; Wang, M.C.; Hung, C.H.; Li, P.R.; Lin, J.J. Antimicrobial activities and cellular responses to natural silicate clays and derivatives modified by cationic alkylamine salts. *ACS Appl. Mater. Interfaces* **2009**, *11*, 2556–2564. [CrossRef] [PubMed]
12. Li, P.R.; Wei, J.C.; Chiu, Y.F.; Su, H.L.; Peng, F.C.; Lin, J.J. Evaluation on cytotoxicity and genotoxicity of the exfoliated silicate nanoclay. *ACS Appl. Mater. Interfaces* **2010**, *2*, 1608–1613. [CrossRef] [PubMed]
13. Wei, J.C.; Yen, Y.T.; Su, H.L.; Lin, J.J. Inhibition of bacterial growth by the exfoliated clays and observation of physical capturing mechanism. *J. Phys. Chem. C* **2011**, *115*, 18770–18775. [CrossRef]

14. Lin, J.J.; Lin, W.C.; Li, S.D.; Lin, C.Y.; Hsu, S.H. Evaluation of the antibacterial activity and biocompatibility for silver nanoparticles immobilized on nanosilicate platelets. *ACS Appl. Mater. Interfaces* **2013**, *5*, 433–443. [CrossRef] [PubMed]
15. Chiao, S.H.; Lin, S.H.; Shen, C.I.; Liao, J.W.; Bau, I.J.; Wei, J.C.; Tseng, L.P.; Hsu, S.H.; Lai, P.S.; Lin, S.Z.; et al. Efficacy and safety of nanohybrids comprising silver nanoparticles and silicate clay for controlling Salmonella infection. *Int. J. Nanomed.* **2012**, *7*, 2421–2432.
16. Su, H.L.; Lin, S.H.; Wei, J.C.; Pao, I.C.; Chiao, S.H.; Huang, C.C.; Lin, S.Z.; Lin, J.J. Novel nanohybrids of silver particles on clay platelets for inhibiting silver-resistant bacteria. *PLoS ONE* **2011**, *6*, e21125. [CrossRef]
17. Liang, J.J.; Wei, J.C.; Lee, Y.L.; Hsu, S.H.; Lin, J.J.; Lin, Y.L. Surfactant-modified nanoclay exhibits an antiviral activity with high potency and broad spectrum. *J. Virol.* **2014**, *88*, 4218–4228. [CrossRef]
18. Ye, S.; Wang, Y.; Jiao, F.; Zhang, H.; Lin, C.; Wu, Y.; Zhang, Q. The role of NADPH oxidase in multi-walled carbon nanotubes-induced oxidative stress and cytotoxicity in human macrophages. *J. Nanosci. Nanotechnol.* **2011**, *11*, 3773–3781. [CrossRef]
19. Kaczmarek, A.; Vandenabeele, P.; Krysko, D.V. Necroptosis: The release of damage-associated molecular patterns and its physiological relevance. *Immunity* **2013**, *38*, 209–223. [CrossRef]
20. Formigli, L.; Papucci, L.; Tani, A.; Schiavone, N.; Tempestini, A.; Orlandini, G.E.; Capaccioli, S.; Orlandini, S.Z. Aponecrosis: Morphological and biochemical exploration of a syncretic process of cell death sharing apoptosis and necrosis. *J. Cell Physiol.* **2000**, *182*, 41–49. [CrossRef]
21. Kusano, H.; Shimizu, S.; Koya, R.C.; Fujita, H.; Kamada, S.; Kuzumaki, N.; Tsujimoto, Y. Human gelsolin prevents apoptosis by inhibiting apoptotic mitochondrial changes via closing VDAC. *Oncogene* **2000**, *19*, 4807–4814. [CrossRef] [PubMed]
22. Luschen, S.; Ussat, S.; Scherer, G.; Kabelitz, D.; Adam-Klages, S. Sensitization to death receptor cytotoxicity by inhibition of Fas-associated death domain protein (FADD)/caspase signaling. *J. Biol. Chem.* **2000**, *275*, 24670–24678. [CrossRef] [PubMed]
23. Delavallee, L.; Cabon, L.; Galan-Malo, P.; Lorenzo, H.K.; Susin, S.A. AIF-mediated caspase-independent necroptosis: A new chance for targeted therapeutics. *IUBMB Life* **2011**, *63*, 221–232. [CrossRef] [PubMed]
24. Lordan, S.; Kennedy, J.E.; Higginbotham, C.L. Cytotoxic effects induced by unmodified and organically modified nanoclays in the human hepatic HepG2 cell line. *J. Appl. Toxicol.* **2011**, *31*, 27–35. [CrossRef]
25. Nel, A.; Xia, T.; Madler, L.; Li, N. Toxic potential of materials at the nanolevel. *Science* **2006**, *311*, 622–627. [CrossRef]
26. Murakami, Y.; Miller, J.W.; Vavvas, D.G. RIP kinase-mediated necrosis as an alternative mechanisms of photoreceptor death. *Oncotarget* **2011**, *2*, 497–509. [CrossRef]
27. Luders, J.; Demand, J.; Höhfeld, J. The ubiquitin-related BAG-1 provides a link between the molecular chaperones Hsc70/Hsp70 and the proteasome. *J. Biol. Chem.* **2000**, *275*, 4613–4617. [CrossRef]
28. Egger, L.; Schneider, L.; Rheme, C.; Tapernoux, M.; Hacki, J.; Borner, C. Serine proteases mediate apoptosis-like cell death and phagocytosis under caspase-inhibiting conditions. *Cell Death Differ.* **2003**, *10*, 1188–1203. [CrossRef]
29. Nabeshi, H.; Yoshikawa, T.; Matsuyama, K.; Nakazato, Y.; Tochigi, S.; Kondoh, S.; Hirai, T.; Akase, T.; Nagano, K.; Abe, Y.; et al. Amorphous nanosilica induce endocytosis-dependent ROS generation and DNA damage in human keratinocytes. *Part. Fibre Toxicol.* **2011**, *8*, 1–10. [CrossRef]
30. Frohlich, E. The role of surface charge in cellular uptake and cytotoxicity of medical nanoparticles. *Int. J. Nanomed.* **2012**, *7*, 5577–5591. [CrossRef]
31. Janer, G.; Fernandez-Rosas, E.; Mas del Molino, E.; Gonzalez-Galvez, D.; Vilar, G.; Lopez-Iglesias, C.; Ermini, V.; Vazquez-Campos, S. In vitro toxicity of functionalized nanoclays is mainly driven by the presence of organic modifiers. *Nanotoxicology* **2014**, *8*, 279–294. [CrossRef] [PubMed]
32. Xie, Y.L.; Pan, Y.E.; Chang, C.J.; Tang, P.C.; Huang, Y.F.; Walzem, R.L.; Chen, S.E. Palmitic acid in chicken granulosa cell death-lipotoxic mechanisms mediate reproductive inefficacy of broiler breeder hens. *Theriogenology* **2012**, *78*, 1917–1928. [CrossRef] [PubMed]
33. Liu, Z.C.; Xie, Y.L.; Chang, C.J.; Su, C.M.; Chen, Y.H.; Huang, S.Y.; Walzem, R.L.; Chen, S.E. Feed intake alters immune cell functions and ovarian infiltration in broiler hens-implications for reproductive performance. *Biol. Reprod.* **2014**, *90*, 134. [CrossRef] [PubMed]
34. Rottenberg, H.; Wu, S. Quantitative assay by flow cytometry of the mitochondrial membrane potential in intact cells. *Biochim. Biophys. Acta* **1998**, *1404*, 393–404. [CrossRef]

35. Chang, L.C.; Chiang, S.K.; Chen, S.E.; Yu, Y.L.; Chou, R.H.; Chang, W.C. Heme oxygenase-1 mediates BAY 11-7085 induced ferroptosis. *Cancer Lett.* **2018**, *416*, 124–137. [CrossRef]
36. Lin, J.-J.; Chang, C.-H.; Lee, B.-H.; Liao, Q.-Y.; Chen, J.-J. Modified Nano Silicate Platelet and Application of Modified Nano silicate Platelet for Preparing Cancer Treating Medicine. ROC Patent I681,777, 16 March 2019.

toxins

MDPI

Article

Polydatin Protects Bovine Mammary Epithelial Cells against Zearalenone-Induced Apoptosis by Inhibiting Oxidative Responses and Endoplasmic Reticulum Stress

Yurong Fu [1,†], Yongcheng Jin [1,†], Anshan Shan [2], Jing Zhang [1,*], Hongyu Tang [1], Jinglin Shen [1], Changhai Zhou [1], Hao Yu [1], Hengtong Fang [1], Yun Zhao [1], Junxiong Wang [1] and Yue Tian [1]

1 Key Laboratory of Zoonosis Research, Department of Animal Science, College of Animal Sciences, Jilin University, ministry of Education, Changchun 130062, China; fuyr20@mails.jlu.edu.cn (Y.F.); ycjin78@163.com (Y.J.); tanghy@jlu.edu.cn (H.T.); shenjinglinshen@aliyun.com (J.S.); zhouch@jlu.edu.cn (C.Z.); yu_hao@jlu.edu.cn (H.Y.); fanght@jlu.edu.cn (H.F.); zhao_yun@jlu.edu.cn (Y.Z.); jxwang19@mails.jlu.edu.cn (J.W.); letian20@mails.jlu.edu.cn (Y.T.)

2 Institute of Animal Nutrition, Northeast Agricultural University, Harbin 150030, China; asshan@neau.edu.cn

* Correspondence: zhang_jing99@jlu.edu.cn; Tel.: +86-147-9436-7799

† YuRong Fu and Yongcheng Jin should be considered joint first author.

Abstract: Zearalenone (ZEA) is a mycotoxin of the *Fusarium* genus that can cause endoplasmic reticulum (ER) stress and Apoptosis in bovine mammary epithelial cells (MAC-T). Polydatin (PD), a glycoside purified from *Polygonum cuspidatum*, has antioxidant properties. This study aimed to explore whether PD can alleviate ZEA-induced damage on bovine mammary epithelial cells (MAC-T). We found that incasing the concentration of ZEA (0, 7.5, 15, 30, 60, 90, 120, and 240 μM) gradually decreased the cell viability. PD treatment alone at 5, 10, and 20 μM did not affect cell viability. Follow-up studies then applied 30 μM of ZEA and 5 μM of PD to treat cells; the results showed that the ZEA + PD treatment group effectively reduced cell oxidative damage compared with the ZEA treatment group. The qPCR analysis showed that ZEA treatment significantly up-regulated the expression of ER stress-related genes, relative to the control. However, adding PD significantly down-regulated the expression of ER stress-related genes. The cell apoptosis detection results showed that, compared with the ZEA treatment group, the ZEA + PD treatment group down-regulated the *Bax* gene and up-regulated the *Bcl-2* gene expressions, which reduced the cell apoptosis rate and Caspase-3 activity. Taken together, these results indicate that PD reduces ZEA-induced apoptosis by inhibiting oxidative damage and ER stress.

Keywords: apoptosis; bovine mammary epithelial cells; endoplasmic reticulum stress; polydatin; zearalenone

Key Contribution: ZEA induces oxidative damage to bovine mammary epithelial cells, leading to various effects. PD can alleviate the oxidative stress response of bovine mammary epithelial cells to ZEA by reducing the subsequent ER stress response.

Citation: Fu, Y.; Jin, Y.; Shan, A.; Zhang, J.; Tang, H.; Shen, J.; Zhou, C.; Yu, H.; Fang, H.; Zhao, Y.; et al. Polydatin Protects Bovine Mammary Epithelial Cells against Zearalenone-Induced Apoptosis by Inhibiting Oxidative Responses and Endoplasmic Reticulum Stress. *Toxins* **2021**, *13*, 121. https://doi.org/10.3390/toxins13020121

Received: 5 January 2021
Accepted: 2 February 2021
Published: 5 February 2021

1. Introduction

Unsuitable storage conditions and variable temperatures can promote the production of mycotoxins in harvested crops. Zearalenone (ZEA) is a fusarium toxin commonly found in feed products. Zearalenone is cytotoxic [1] to the liver, spleen [2], intestines [3], and hematopoietic cells [4]. It induces the production of reactive oxygen species (ROS) and triggers intracellular oxidative stress, which has subsequent cytotoxic and genotoxic effects [5]. Furthermore, oxidative damage reduces the performance of dairy cows and affects their disease resistance, which reduces milk yield and quality [6]. Oxidative stress is closely related to the cow's physiology and nutritional status factors. Excessive ROS production leads to oxidative stress, loss of cellular function, and, ultimately, apoptosis or necrosis [7,8]. Our previous studies have shown that adding ZEA to bovine mammary epithelial cells promotes excessive ROS generation, endoplasmic reticulum (ER) stress, and apoptosis [9]. The balance between reduced glutathione (GSH) and oxidized glutathione (GSSH) is essential for maintaining cellular redox homeostasis. However, GSH gets depleted during protein misfolding, resulting in ROS generation [10]. Various parameters, such as malondialdehyde (MDA), reduced glutathione (GSH), total superoxide dismutase (T-SOD), and antioxidant enzyme activity, among others, are related to oxidative stress, and can indicate the cellular levels of oxidative stress [11].

ZEA was previously detected in milk samples. Therefore, whether ZEA may have similar effects on bovine mammary epithelial cells is worth studying. Studies have shown that ZEA induces mouse Leydig cell apoptosis by activating the ER stress-dependent signaling pathway [12]. The function of the ER is sensitive to the accumulation of unfolded proteins, calcium homeostasis, and redox state changes; thus, disruption of these processes can cause ER stress. Since the unfolded protein response (UPR) is an ER stress response that maintains cell homeostasis, deregulation of UPR can cause cell disorders and cell death [13]. However, previous studies found that ZEA can induce apoptosis of MAC-T cells. Mammalian cells express the UPR transducer proteins IRE1, PERK, and ATF6, which control transcriptional and translational responses to ER stress [14]. Under non-stress or physiological conditions, these proteins remain in an inactive state, and bind to the molecular chaperone BiP/GRP78, which is also the primary regulator of ER stress. Under ER stress, GRP78 decomposes from the stress sensors, thus activating them. Once activated, PERK phosphorylates the α subunit of the translation initiation factor eIF2α (eukaryotic initiation factor 2α). This activation results in an overall reduction of translation, and promotes priority translation of UPR-dependent genes (such as activated transcription factor 4 (ATF4)). The vital target of ATF4 is CHOP (C/EBP homologous protein) [15].

Oxidative stress and ROS generation are indispensable components of UPR. Therefore, ROS generation can occur both upstream and downstream of UPR. Both oxidative and ER stresses are involved in various physiological and pathophysiological conditions, and they play vital roles in cell homeostasis and apoptosis. This fact suggests that ER stress and oxidative stress have a significant correlation. For instance, our previous studies have shown that ZEA can cause ER stress and apoptosis in MAC-T cells. However, whether there are compounds that can alleviate the effects of ZEA remain unclear.

Polydatin (3,4′,5-trihydroxystibene-3-β-mono-D-glucoside; PD) is a natural precursor of resveratrol and an active ingredient in various Chinese medicines, such as *Polygonum cuspidatum*. Polydatin is the product of resveratrol and glucose, hence, it is also called resveratrol glycoside. Its pharmacological effects are similar to resveratrol, and they can be transformed into each other in the body. However, PD has a stronger antioxidant effect and stability [16]. PD may also exist in peanuts, grapes, and wine [17]. It has a wide range of positive health effects, including antioxidant [18], anti-cardiovascular diseases [19], anti-inflammatory [20], and inhibition of cancer cell growth [16], and can be used for the treatment of shock [21]. Polydatin can protect bone marrow stem cells from oxidative damage [22]; however, it is unclear whether PD can rescue cells from ER stress and apoptosis. Here, we investigated whether PD protects bovine mammary epithelial cells from ZEA-induced ER stress and apoptosis. In this study, various *in vitro* experiments were conducted to determine whether the PD-related effect caused by ZEA on MAC-T cells can be alleviated.

2. Results

2.1. The Increase in PD Decreased the MAC-T Cell Viability Due to ZEA

Cell viability was measured by the CCK-8 assay. We found that ZEA exposure significantly reduced the MAC-T cell viability, and viability decreased with increasing ZEA concentrations (Figure 1a). Cells exposed to 30 μM of ZEA had significantly lower survival than the control group ($p < 0.001$). Therefore, cells were treated with 30 μM of ZEA in subsequent experiments. MAC-T cells treated with 5–20 μM of PD had similar survival rates as the control group, suggesting that PD does not affect cell survival (Figure 1b). Given that all concentrations yielded similar results, 5 μM of PD was used for subsequent experiments. Interestingly, MAC-T cells exposed to ZEA + PD had significantly higher survival rates than cells treated with ZEA alone ($p < 0.01$; Figure 1c). The cell viability of the ZEA treatment group decreased by about 13% relative to the control group. Compared with the ZEA treatment group, the ZEA + PD treatment group increased by about 9%. The addition of PD effectively increased cell viability due to ZEA inhibition.

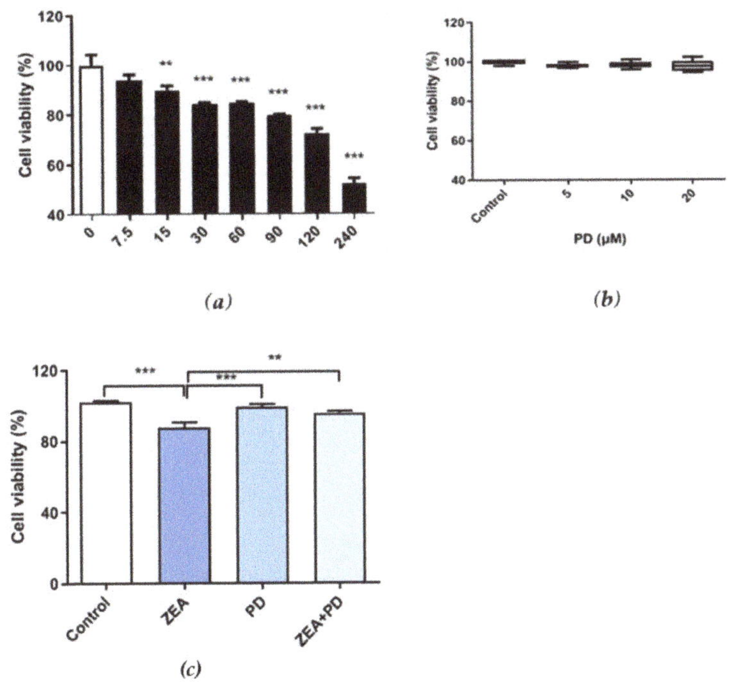

Figure 1. Effects of Zearalenone (ZEA) and Polydatin (PD) on the viability of bovine mammary epithelial cells (MAC-T). (**a**) Viability of MAC-T cells treated with different concentrations (0, 7.5, 15, 30, 60, 90, 120, and 240 μM) of ZEA for 24 h. (**b**) Viability of MAC-T cells treated with different concentrations (0, 5, 10, and 20 μM) of PD for 24 h. (**c**) Effects of 30 μM of ZEA and 5 μM of PD on MAC-T cell viability. Figure 1c shows the control group, ZEA treatment group, PD treatment group, and ZEA + PD treatment group. Each experiment was repeated three times. All values are expressed as mean ± SEM ($n = 3$). In the figure, ** $p < 0.01$, *** $p < 0.001$.

2.2. *PD Reduces the ZEA-Induced Cytotoxic Effect in MAC-T Cells*

The lactate dehydrogenase (LDH) activity, measured by quantifying the LDH levels released from disrupted cells into the culture media, can serve as a proxy for cytotoxicity. The results showed that LDH increased with increasing ZEA concentrations (Figure 2a). Compared with the ZEA-treated group, the LDH content of cells treated with ZEA + PD was significantly reduced (Figure 2b, $p < 0.001$). Meanwhile, the LDH content of the ZEA treatment group increased by about 43% relative to the control group. Compared with the ZEA treatment group, the ZEA + PD treatment group decreased by about 23%. These results suggest that PD lessens the cytotoxic effect in MAC-T cells treated with ZEA due to ZEA and PD interactions.

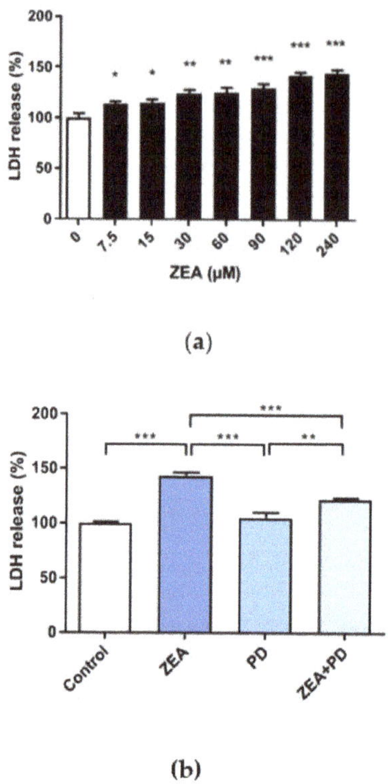

Figure 2. Effects of ZEA and PD on the lactose dehydrogenase (LDH) levels of MAC-T cells. (**a**) LDH released by MAC-T cells treated with varying ZEA concentrations (0, 7.5, 15, 30, 60, 90, 120, and 240 μM) for 24 h. (**b**) LDH released by MAC-T cells exposed to 30 μM of ZEA and 5 μM of PD. In the figure, * $p < 0.05$, ** $p < 0.01$, *** $p < 0.001$.

2.3. *PD Can Inhibit ZEA-Induced Oxidative Damage*

Previous results indicate that ZEA and PD can increase cell viability and reduce the LDH activity of treated cells. Therefore, whether the effect of oxidative damage is obvious requires further investigation. This study showed that the MDA level of the ZEA treatment group was significantly higher than that of the control group ($p < 0.01$, Figure 3a). In contrast, the MDA level of the ZEA + PD group was significantly lower than the ZEA group ($p < 0.001$). Compared with the control group, the MDA level of the ZEA group increased by 1.44 times. Meanwhile, the MDA level of the ZEA + PD treatment group decreased by 1.92 times relative to the ZEA group. The T-SOD content of cells exposed to ZEA was significantly lower than that of the control group ($p < 0.01$, Figure 3b). However, compared with ZEA treatment alone, the ZEA + PD treatment had significantly increased T-SOD content ($p < 0.01$). Finally, we evaluated the GSH content; we found that the GSH content of the ZEA treatment group was significantly lower than the control group ($p < 0.001$). Compared with the control group, the T-SOD activity of the ZEA treatment group decreased by 1.14 times. On the contrary, the GSH level of the ZEA + PD treatment group was significantly higher than that of the ZEA treatment group ($p < 0.05$). These results indicate that PD can inhibit the oxidative damage caused by ZEA in MAC-T cells.

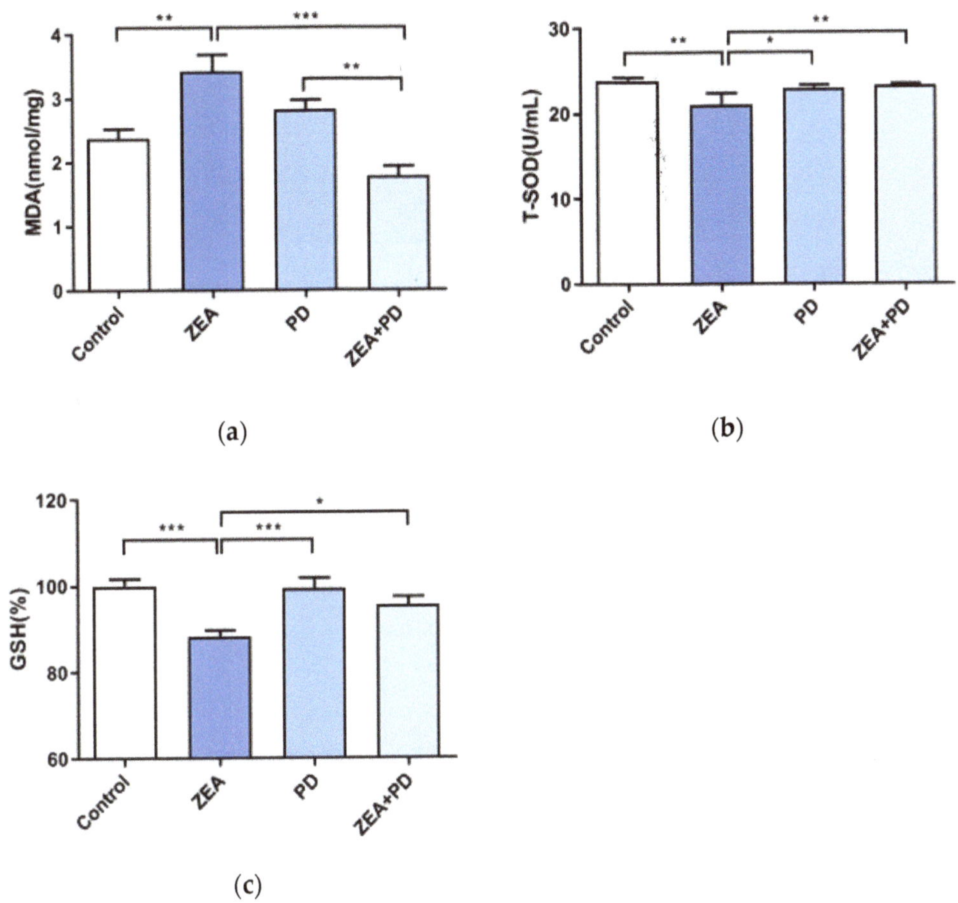

Figure 3. PD inhibits ZEA-induced oxidative damage in MAC-T cells. (**a**) MDA content, (**b**) T-SOD content, and (**c**) GSH content. All values are expressed as mean ± SEM ($n = 3$). In the figure, * means $p < 0.05$, ** means $p < 0.01$, and *** means $p < 0.001$.

2.4. PD Inhibits ZEA-Induced Increases in ROS Levels

The above studies indicate that ZEA causes oxidative damage to MAC-T cells, and that PD can alleviate this situation. ROS mediates oxidative stress, and is a necessary marker for detecting ROS. The standard ROS detection system uses the fluorescent probe DCFH-DA. In this study, the ZEA treatment group had significantly higher active oxygen content than the control group ($p < 0.001$; Figure 4a,b), and the ZEA + PD combination treatment significantly lowered the active oxygen content induced by the ZEA treatment ($p < 0.001$). While ZEA increased the ROS level by 2.94 times, adding PD reduced it by 1.97 times.

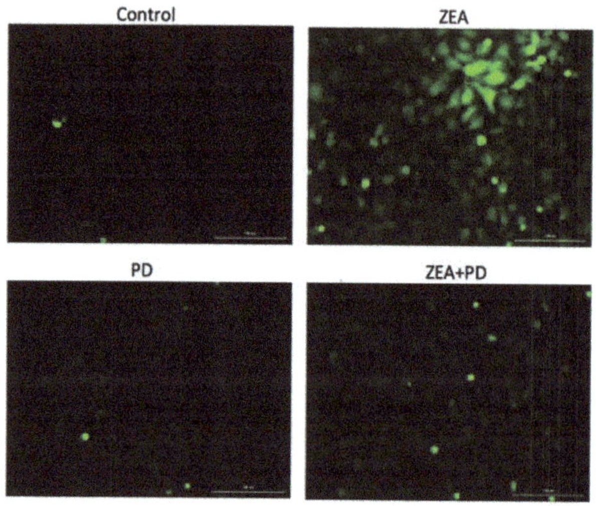

(a)

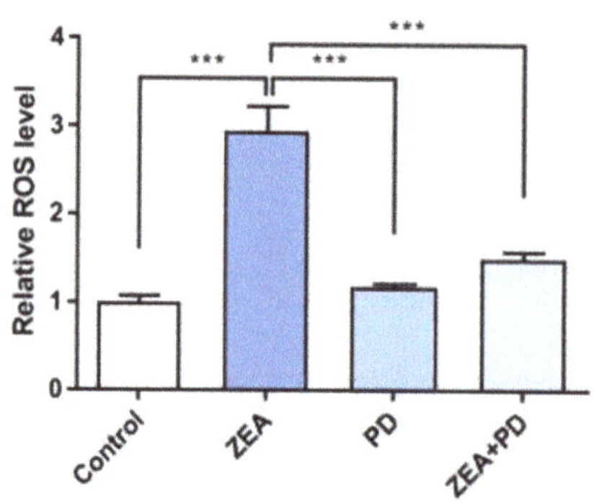

(b)

Figure 4. The effect of PD on ZEA-induced production of ROS in MAC-T cells. (**a**) MAC-T cells from the control, ZEA, PD, and ZEA + PD treatment groups stained with the DCFH-DA probe, and (**b**) their fluorescence intensity. In the figure, *** means $p < 0.001$. Scale bar = 200 μM.

2.5. PD Inhibits ZEA-Induced ER Stress in MAC-T Cells

Both oxidative and ER stress participate in various physiological and pathophysiological conditions, and play a vital role in cell homeostasis and apoptosis. This study measured the expression levels of genes associated with ER stress to understand whether exposure to ZEA damages the ER function. The expressions of the ER stress markers *GRP78* ($p < 0.05$), *ATF4* ($p < 0.01$), *ATF6* ($p < 0.05$), *ASK1* ($p < 0.05$), and *CHOP* ($p < 0.01$) were significantly up-regulated in the ZEA treated group relative to the control group (Figure 5a–e). However, the expression levels of *GRP78* ($p < 0.05$), *ATF4* ($p < 0.001$), *ATF6* ($p < 0.05$), and *CHOP* ($p < 0.01$) in the ZEA + PD treatment group were significantly lower than the ZEA treatment group. *ASK1* expression showed a decreasing trend in the ZEA + PD group compared to the ZEA group. ZEA significantly increased the expressions of the ER stress marker genes by more than two-fold, but returned to normal levels after adding PD.

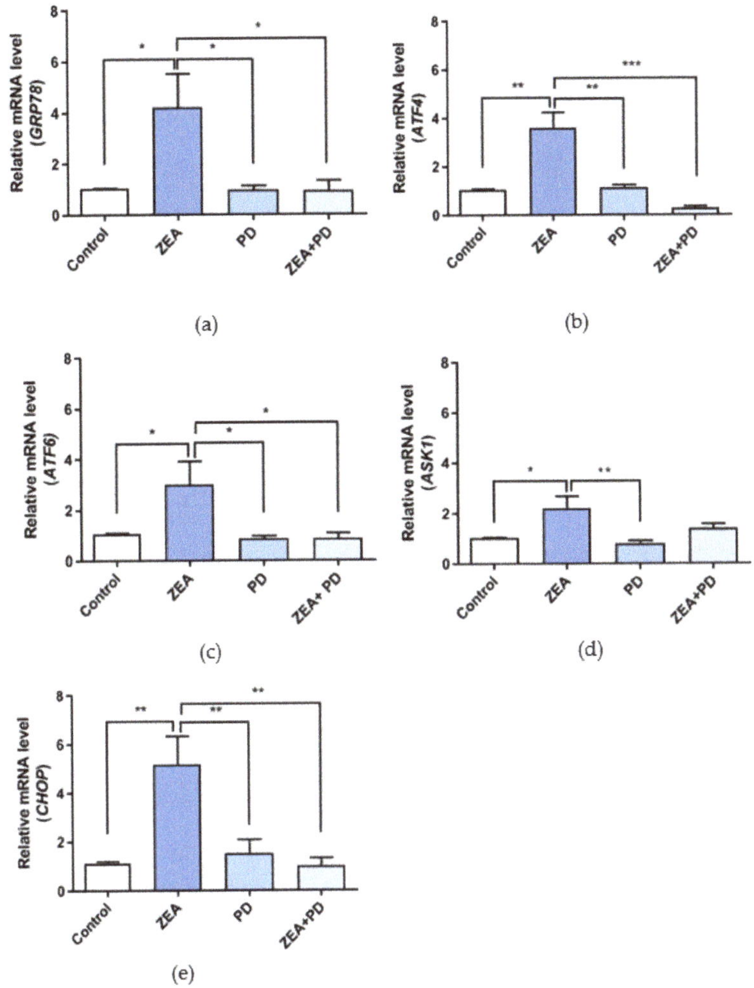

Figure 5. The effect of PD on the expressions of ER stress marker genes in MAC-T cells treated with by ZEA. (**a**–**e**) Analysis of gene expression levels of *GRP78*, *ATF4*, *ATF6*, *ASK1*, and *CHOP*. In the figure, * means $p < 0.05$, ** means $p < 0.01$, and *** means $p < 0.001$.

2.6. PD Inhibits ZEA-Induced Apoptosis

Compared to the control, treating MAC-T cells with ZEA increased the apoptotic cell rate ($p < 0.001$, Figure 6a,b); it increased from 10.22% to 29.67%. The ZEA + PD treatment group had a significantly ($p < 0.01$) lower apoptotic rate (29.67%) than the ZEA treatment group (29.06%). *Bax* expression was significantly up-regulated ($p < 0.01$), while *Bcl-2* expression was significantly down-regulated ($p < 0.05$) in the ZEA-treated group relative to the control group (Figure 6c,d). In contrast, the ZEA + PD group had significantly lower *Bax* expression ($p < 0.001$) and significantly higher *Bcl-2* expression ($p < 0.05$) than ZEA alone. Finally, we measured the Caspase-3 activity using a Caspase-3 activity assay; the Caspase-3 activity of the ZEA-treated group was significantly higher than the control group ($p < 0.05$, Figure 6e). However, the Caspase-3 activity of the ZEA + PD treatment group was significantly lower than ZEA alone ($p < 0.01$). Taken together, these results suggest that treating MAC-T cells with PD reduces the apoptotic effects of ZEA.

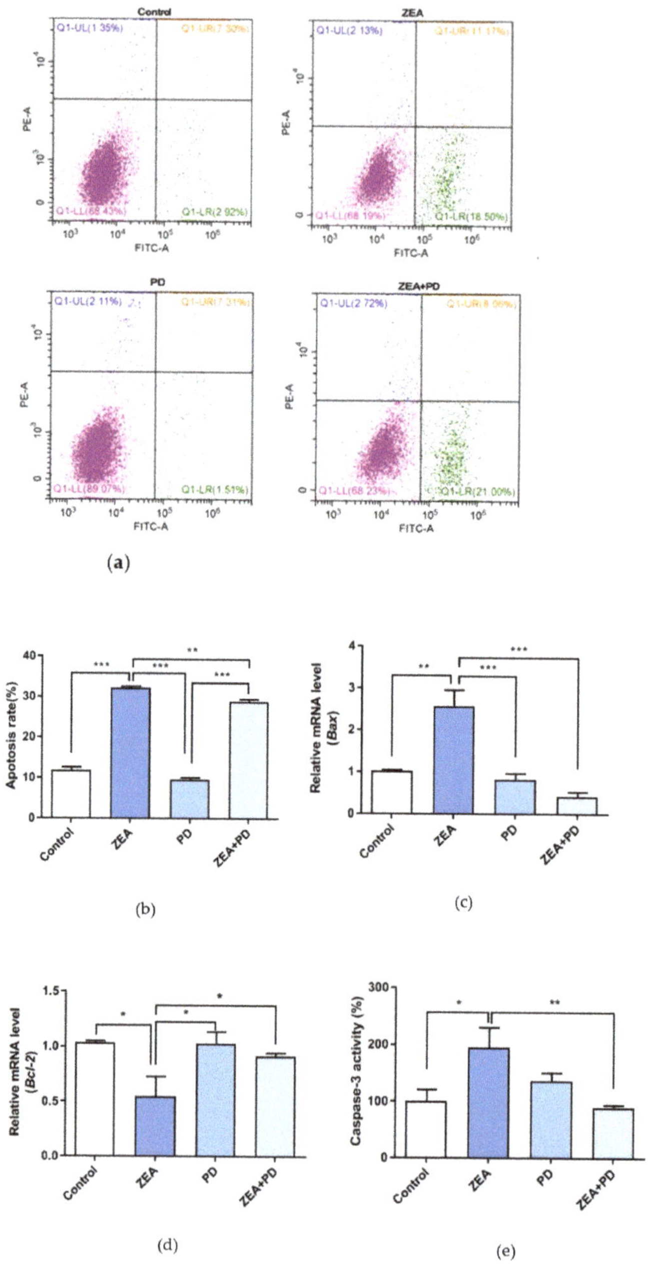

Figure 6. The effect of PD on ZEA-induced apoptosis in MAC-T cells. (**a**) Apoptosis measurements using annexin V/PI. The lower left panel contains annexin V and PI negative cells, while the lower right panel is annexin V positive. The upper left panel is annexin V− and PI+, while the upper right panel is annexin V+ and PI+. (**b**) Four groups of apoptotic rates. (**c**,**d**) qPCR analysis of mRNA expression levels of apoptosis-related genes *Bcl-2* and *Bax*. (**e**) Four groups of Caspase-3 activity. In the figure, * means $p < 0.05$, ** means $p < 0.01$, and *** means $p < 0.001$.

3. Discussion

Feeding cows with a moldy feed containing mycotoxins results in a lower feed intake and reduced milk production [23]. In addition, the harmful mycotoxins in feed can be secreted in milk [24]. Currently, most countries have clear guidelines and detection limits for milk toxins. Although ZEA was detected in milk, the content may not be harmful to the human body. However, for the dairy cow itself, after feeding with the feed containing ZEA, if the ZEA is transferred to the milk through the bovine mammary gland, it cannot be ignored that ZEA is harmful. Therefore, it is necessary to evaluate the toxic effects of ZEA on breast cells at the cellular level and seek ways to protect breast cells from this damage, which is beneficial for the subsequent impact on milk production.

Zearalenone, a major toxin of animal feed, deserves attention. In our previous research, MAC-T cells were used as a model to study the effects of ZEA. It was found that ZEA can inhibit cell viability, reduce mitochondrial membrane potential, and cause ER stress and apoptosis [9]. In this study, PD, a protective agent, alleviated the effect of ZEA on bovine mammary epithelial cells. We found that ZEA decreased the viability of bovine mammary epithelial cells in a dose-dependent manner, consistent with our previous findings [9]. In another study, treating human hepatoma cells (HepG2 cells) with 50 to 250 μM of ZEA for 24 h significantly reduced cell viability [25].

Recent pharmacological studies have shown that *Polygonum cuspidatum* has antibacterial, anti-inflammatory, diuretic, and other effects [26]. One of its main ingredients, PD, has various biological activities and pharmacological effects that are often cell type specific. Additionally, PD has anti-inflammatory and antioxidant properties, and can effectively treat health conditions and diseases [27]. Furthermore, PD is reported to improve arsenic damage to rat testicular cells [28] and inhibit the growth of some human tumor cells [29]. The cytotoxicity of a compound can be measured by the LDH levels released by cells [26]. Generally, an increase in LDH levels indicates increased cytotoxicity. However, adding PD significantly reduced LDH release by MAC-T cells exposed to ZEA.

SOD is an antioxidant enzyme, and its activity indirectly reflects the body's ability to eliminate free radicals. MDA is the end product of lipid peroxidation, and the MDA levels reflect the damage caused by lipid peroxidation [16]. Measuring the levels of these compounds can provide insights into the levels of cellular oxidative damage. ZEA treatment increased the MDA content and decreased GSH and T-SOD levels, suggesting that ZEA increases oxidative damage in cells. In another study, ZEA significantly increased the Glutathione peroxidase (GPx), Catalase (CAT), and SOD activities in the testis tissue of adult Balb/c male mice, but significantly increased the MDA in the same tissue [30]. However, all of the tested concentrations of PD, including the lowest, alleviated the oxidative damage caused by ZEA on MAC-T cells. Exposing cells to both ZEA and PD reduced the MDA content and increased both GSH and T-SOD contents [28,31]. Studies have shown that PD can reduce oxidative damage in cardiomyocytes [19], consistent with the finding of the present study, where PD reduced ZEA-induced oxidative damage of MAC-T cells.

ER stress can activate mitochondrial pathways that trigger apoptosis, a reaction characterized by increased ROS production and lipid peroxidation, loss of mitochondrial transmembrane potential, activation of cysteine and acid proteases, and DNA damage [32,33]. We found that exposure to ZEA increased the ROS levels in MAC-T cells. It has previously been reported that ROS is produced in ZEA-treated leukemia cells (HL-60) [34]. However, treating cells with PD and ZEA reduced the ROS levels relative to the ZEA-treated group. Furthermore, PD can inhibit the damage induced by *S. aureus* lipoteichoic acid by attenuating ROS production [35]. These results concur with our findings that PD can effectively reduce the ROS contents of MAC-T cells.

GRP78 and *CHOP* are markers of ER stress [36], and the up-regulation of these genes indicates increased ER stress. We found that exposing MAC-T cells to ZEA significantly up-regulated these genes, as well as *ATF4*, *ATF6*, and *ASK1*. These results are consistent with previous studies [9]. However, treating ZEA-exposed cells with PD reduced the expression levels of *GRP78*, *CHOP*, *ATF4*, and *ATF6* compared with the ZEA treatment group. These results show that PD can relieve the ZEA-induced ER stress on MAC-T cells.

The CCAAT/enhancer-binding protein homolog protein (CHOP) is a key pro-apoptotic transcription factor related to ER stress [32]. It is inadvertently known that the *Bcl-2* gene is anti-apoptotic, while the *Bax* gene is pro-apoptotic. CHOP-mediated ER stress inhibits *Bcl-2* expression, promoting *Bax* expression to induce apoptosis [37,38]. We found that cells exposed to ZEA had lower *Bcl-2* expression and elevated *Bax* expression. However, adding PD increased *Bcl-2* expression and decreased *Bax* expression. Flow cytometry evaluation of the apoptotic rate showed that ZEA treatment significantly increased apoptosis compared to the control group; however, the addition of PD significantly reduced the apoptosis rate. Although PD significantly reduced the cell apoptosis rate, the reduction degree was not large, indicating that 5 μM of PD can ameliorate cell apoptosis, but not completely inhibit it. Banjerdpongchai et al. demonstrated that ZEA could activate Caspase-3 activity in HL-60 and U937 cells in a dose-dependent manner [34]. In addition, Caspase-3 activity was significantly higher in ZEA-treated cells, and significantly decreased in the ZEA + PD treatment group compared to cells treated with ZEA. Our results are consistent with the anti-apoptotic properties of PD demonstrated by Liu et al. [39]. Taken together, our findings suggest that PD can attenuate ZEA-induced ER stress and apoptosis.

These results show that PD effectively reduces oxidative stress, ER stress, and apoptosis induced by ZEA; it may be a natural protective agent against ZEA. Therefore, this mechanism of PD should be studied further.

4. Conclusions

In vitro experiments revealed a new phenomenon that PD reduces ZEA-induced oxidative damage and ER stress in MAC-T cells by reducing ROS production, the activity of antioxidant enzymes, and the expression of ER stress-related genes. The stress subsequently alleviates cell apoptosis. These findings indicate that PD could be an effective antioxidant and potential therapeutic agent for diseases related to oxidative and ER stresses. However, the possible molecular mechanism of PD's protective effect on ZEA-induced cell damage in MAC-T cells needs further research. Thus, *in vivo* studies are required to determine the role of PD as an effective therapeutic agent that can be used as a feed additive.

5. Materials and Methods

5.1. Chemicals and Reagents

Zearalenone (ZEA, purity >99%), hydrocortisone, penicillin-streptomycin, and insulin were purchased from Sigma-Aldrich (St. Louis, MO, USA). ZEA was dissolved in dimethyl sulfoxide (DMSO, Sigma Chemical Co., St. Louis, MO, USA) and stored at −20 °C. Polydatin (PD) was purchased from Refinsen Biotech Co., Ltd. (Chengdu, China), and dissolved in DMSO to generate a 100 mM stock solution, which was stored at −20 °C. Fetal bovine serum (FBS) was purchased from Gibco (Gaithersburg, MD, USA), while Dulbecco's modified Eagle's/high-glucose medium (DMEM) was purchased from Hyclone (Logan, UT, USA). The cell counting kit-8 (CCK8) was procured from Dojindo Laboratories (Kumamoto, Japan), while the FITC Annexin V apoptosis detection kit was purchased from BD Biosciences (San Jose, CA, USA). Kits to measure ROS and caspase-3 activity were obtained from Beyotime Biotechnology (Shanghai, China). Kits for detecting the T-SOD, MDA, GSH, and lactate dehydrogenase (LDH) activities were purchased from Jiancheng Bioengineering Institute (Nanjing, China). Reagents for qPCR applications included SYBR green for real-time PCR (TransGen Biotech, Beijing, China) and the RevertAid First Strand cDNA Synthesis Kit (CW0581, Beijing, China).

5.2. Cell Culture

The bovine mammary epithelial cell line MAC-T was kindly provided by Professor Hong Gu Lee (Konkuk University, Seoul, Korea). For *in vitro* analyses, MAC-T cells were maintained in DMEM/high-glucose media containing 10% FBS, 1% penicillin-streptomycin, 1 μg/mL of hydrocortisone, and 5 μg/mL of insulin, kept in a 37 °C incubator with 5% CO_2.

5.3. Cell Viability Assay

Cell viability was measured using a CCK-8 kit (Dojindo Laboratories, Kumamoto, Japan) following the manufacturer's instructions. Briefly, MAC-T cells were seeded in 96-well plates at a density of 1×10^4 cells/well. When the cells reached 70–80% confluence, they were treated with different concentrations of ZEA (0, 7.5, 15, 30, 60, 90, 120, and 240 μM) and PD (0, 5, 10, and 20 μM) for 24 h. A cell viability test was then performed; 10 μL of CCK8 reagent were added to each well, and the cells were incubated for an additional 2.5 h at 37 °C, then randomly divided into four groups, each with three repeats. The experiments included the control, ZEA, PD, and ZEA + PD treatment groups. The absorbance at 450 nm was measured using a microplate reader (Eon, BioTek Instruments, USA), and cell viability was calculated as follows: (Treatment Group OD–Blank Group OD)/(Control Group OD–Blank Group OD).

5.4. Lactate Dehydrogenase Assay

We used an LDH kit (Nanjing Jiancheng Bioengineering Institute, Nanjing, China) to detect the LDH activity by the microplate method. The cells were plated in 96-well plates; ZEA and PD were treated when the cells' confluence reached 70–80%. After 24 h, the cell culture supernatant was collected for cytotoxicity level detection. Finally, the reagents were added according to the kit instructions, and the absorbance was measured with a microplate reader at a wavelength of 450 nm.

5.5. Detection of GSH, T-SOD, and MDA Levels

The levels of GSH, T-SOD, and MDA were tested using the respective assay kits and following the manufacturer's protocol. As described above, MAC-T cells were treated with varying concentrations of ZEA and PD for 24 h, and cells were collected to measure oxidation levels. The culture solution was also collected for later use. The cells were taken out with a cell scraper and transferred to a 1.5 mL centrifuge tube. Then, 500 μL of the extract was added, and the contents were mixed by homogenization. Subsequently, 100 uL of the mixture were transferred to another 1.5 mL centrifuge tube. The BCA kit determines the protein concentration. We measured the absorbance at 530 nm in the microplate reader. To determine the T-SOD content, the cells were cultivated in a similar way, then the cell protein concentration was evaluated. The kit instructions were used to add the reagents, which were mixed well and kept at room temperature for 10 min, then the wavelength of 550 nm was colorimetrically detected. To determine the GSH content, cultured cells were taken out by cell scraping and transferred to a 1.5 mL centrifuge tube. A glass homogenizer was used for mixing; 100 μL of the precipitation solution was taken and centrifuged at 3500 rpm for 10 min, and then the supernatant was taken for detection. The absorbance was measured at 405 nm in the microplate reader.

5.6. Measurement of ROS Production

The intracellular ROS levels of MAC-T cells were measured by a DCFH-DA kit (Shanghai, China). Cultures in multiple six-well plates were selected for the experiment. When the cell confluence reached 70–80%, they were treated and cultured in a medium containing either ZEA or PD for 24 h. Then, cultures were divided into control, ZEA, PD, and ZEA + PD treatment groups, and stained with 10 μL of DCFH-DA for 30 min at 37 °C in the dark. Cells were washed using 1× PBS to remove the unincorporated dye. The green fluorescence intensity was measured using the fluorescence microscope function of a Cytation five-cell imaging reader (BioTek Instruments, Winooski, VT, USA). Data were analyzed using the Gen5 3.03 software (BioTek Instruments, Winooski, VT, USA).

5.7. qPCR

The total RNA was isolated using a TRIzol reagent (Invitrogen, Carlsbad, CA, USA). Complementary (cDNA) was synthesized from 1 μg of RNA at 42 °C for 15 min, and then at 85 °C for 5 min. The SYBR Green Mix Kit was used to perform qPCR reactions. The qPCR reaction program was 94 °C for 30 s, 94 °C for 5 s, 60 °C for 15 s, and 72 °C for 10 s, respectively. A total of 40 cycles were performed. The mRNA levels were calculated by the $2^{-\Delta\Delta Ct}$ comparative method and analyzed by normalization with β-actin mRNA expression. The primers used in the qPCR analyses are listed in Table 1.

Table 1. Gene name and PCR primer sequences.

Gene	Forward Primer	Reverse Primer	GenBank Accession No.	Product Size (bp)
β-actin	5′-CCCTGGAGAAGAGCTACGAG-3′	5′-GTAGTTTCGTGAATGCCGCAG-3′	nm_173979.3	130
GRP78	5′-CGACCCCTGACGAAAGACAA-3′	5′-AGGTGTCAGGCGATTTTGGT-3′	nm_001075148.1	198
ATF4	5′-AGATGACCTGGAAACCATGC-3′	5′-AGGGGAAGAGGTTGAAAGA-3′	nm_001034342.2	190
ATF6	5′-ATATTCCTCCGCCTCCCTGT-3′	5′-GTCCTTTCCACTTCGTGCCT-3′	XM_024989876.1	103
ASK1	5′-GCTATGGAAAGGCAGCCAGA-3′	5′-TCTGCTGACATGGACTCTGG-3′	nm_001144081.2	160
CHOP	5′-GAGCTGGAAGCCTGGTATGA-3′	5′-CTCCTTGTTTCCAGGGGGTG-3′	nm_001078163.1	90
Bax	5′-GCTCTGAGCAGATCATGAAGAC-3′	5′-CAATTCATCTCCGATGCGCT-3′	nm_173894.1	167
Bcl-2	5′-GATGACCGAGTACCTGAACC-3′	5′-AGAGACAGCCAGGAGAAATCA-3′	nm_001166486.1	123

5.8. Flow Cytometry Detection of Apoptosis

Apoptosis was determined by staining cells with annexin V and propidium iodide (PI) using the Annexin V-FITC Apoptosis Detection Kit I (San Jose, CA, USA). MAC-T cells were first seeded in six-well plates and treated with ZEA and/or PD, as described above. After 24 h, all of the cells and culture fluids were collected for subsequent experiments. At the time of sampling, we applied trypsin without EDTA to digest the cells, and then proceeded to the cell wash step. The cells were resuspended with pre-chilled 1*PBS (4 °C) and centrifuged at 1000 rpm for 5 min. This step was repeated thrice (trypsin was washed clean). After discarding the supernatants, cells were centrifuged, resuspended, and incubated for 15 min in 1× annexin binding buffer, 5 μL of FITC Annexin V, and 5 μL of PI working solutions. After incubation, cell apoptosis was detected by flow cytometry (Beckman-Coulter, Shanghai, China).

5.9. Caspase-3 Activity Assay

The caspase-3 activity assay kit (Beyotime Biotechnology, Shanghai, China) was used to measure the caspase-3 activity levels following the manufacturer's protocols. Briefly, cells were trypsinized and centrifuged at 600× *g* for 5 min. The supernatant was discarded, and the pellet was washed with 1× PBS and centrifuged at 600× *g* for 5 min. The pellet was mixed with 100 μL of lysate reagent to resuspend cells, and the mixture was incubated for 15 min on an ice bath. After 15 min of centrifugation at 16,000× *g* and 4 °C, the supernatant was transferred to a pre-chilled centrifuge tube, and the absorbance was measured at 405 nm using a microplate reader (BioTek Instruments, Winooski, VT, USA).

5.10. Statistical Analyses

The data were analyzed using the SPSS statistical software package. All experiments were repeated with three independent replicates. Statistical differences among the treatment groups were calculated using one-way ANOVA, and Duncan's Multiple Range Test was used for multiple comparisons. Data are expressed as mean ± SEM. Differences of $p < 0.05$ were considered statistically significant.

Author Contributions: Conceptualization, Y.J. and J.Z.; Formal analysis, Y.Z.; Funding acquisition, J.Z.; Investigation, J.W. and Y.T.; Methodology, H.Y.; Project administration, A.S. and J.S.; Supervision, C.Z. and H.T.; Visualization, H.F. and H.Y.; Writing—original draft, Y.F.; Writing—review & editing, Y.F. and Y.J. All authors have read and agreed to the published version of the manuscript.

Funding: This research was funded by the Scientific Research Project of Education Department of Jilin Province(JJKH20211133KJ), the Jilin Scientific and Technological Development Program (20190301034NY), P.R. China.

Institutional Review Board Statement: Not applicable.

Informed Consent Statement: Not applicable.

Data Availability Statement: Data sharing not applicable.

Acknowledgments: The MAC-T cell line was a gift from the laboratory of Professor Lee Hong-Gu, Konkuk University, Korea.

Conflicts of Interest: The authors declare that there is no conflict of interest with any commercial or financial organization regarding the material discussed in the manuscript.

References

1. Kouadio, J.H.; Mobio, T.A.; Baudrimont, I.; Moukha, S.; Dano, S.D.; Creppy, E.E. Comparative study of cytotoxicity and oxidative stress induced by deoxynivalenol, zearalenone or fumonisin B1 in human intestinal cell line Caco-2. *Toxicology* **2005**, *213*, 56–65. [CrossRef] [PubMed]
2. Tiemann, U.; Brussow, K.P.; Dannenberger, D.; Jonas, L.; Pohland, R.; Jager, K.; Danicke, S.; Hagemann, E. The effect of feeding a diet naturally contaminated with deoxynivalenol (DON) and zearalenone (ZON) on the spleen and liver of sow and fetus from day 35 to 70 of gestation. *Toxicol. Lett.* **2008**, *179*, 113–117. [CrossRef]
3. Liu, M.; Gao, R.; Meng, Q.; Zhang, Y.; Bi, C.; Shan, A. Toxic effects of maternal zearalenone exposure on intestinal oxidative stress, barrier function, immunological and morphological changes in rats. *PLoS ONE* **2014**, *9*, e106412. [CrossRef]
4. Yang, L.; Yang, W.; Feng, Q.; Huang, L.; Zhang, G.; Liu, F.; Jiang, S.; Yang, Z. Effects of purified zearalenone on selected immunological measurements of blood in post-weaning gilts. *Anim. Nutr.* **2016**, *2*, 142–148. [CrossRef]
5. Pietsch, C.; Noser, J.; Wettstein, F.E.; Burkhardt-Holm, P. Unraveling the mechanisms involved in zearalenone-mediated toxicity in permanent fish cell cultures. *Toxicon Off. J. Int. Soc. Toxinology* **2014**, *88*, 44–61. [CrossRef]
6. Sordillo, L.M.; Aitken, S.L. Impact of oxidative stress on the health and immune function of dairy cattle. *Vet. Immunol. Immunopathol.* **2009**, *128*, 104–109. [CrossRef]
7. Wang, C.C.; Chiang, Y.M.; Sung, S.C.; Hsu, Y.L.; Chang, J.K.; Kuo, P.L. Plumbagin induces cell cycle arrest and apoptosis through reactive oxygen species/c-Jun N-terminal kinase pathways in human melanoma A375.S2 cells. *Cancer Lett.* **2008**, *259*, 82–98. [CrossRef] [PubMed]
8. Moon, D.O.; Kim, M.O.; Kang, S.H.; Choi, Y.H.; Kim, G.Y. Sulforaphane suppresses TNF-alpha-mediated activation of NF-kappaB and induces apoptosis through activation of reactive oxygen species-dependent caspase-3. *Cancer Lett.* **2009**, *274*, 132–142. [CrossRef] [PubMed]
9. Fu, Y.R.; Jin, Y.C.; Zhao, Y.; Shan, A.; Fang, H.T.; Shen, J.L.; Zhou, C.H.; Yu, H.; Zhou, Y.F.; Wang, X.; et al. Zearalenone induces apoptosis inbovine mammary epithelial cells by activating endoplasmic reticulum stress. *J. Dairy Sci.* **2019**, *102*, 10543–10553. [CrossRef] [PubMed]
10. Santos, C.X.; Tanaka, L.Y.; Wosniak, J.; Laurindo, F.R. Mechanisms and implications of reactive oxygen species generation during the unfolded protein response: Roles of endoplasmic reticulum oxidoreductases, mitochondrial electron transport, and NADPH oxidase. *Antioxid. Redox Signal.* **2009**, *11*, 2409–2427. [CrossRef]
11. Samarghandian, S.; Azimi-Nezhad, M.; Farkhondeh, T.; Samini, F. Anti-oxidative effects of curcumin on immobilization-induced oxidative stress in rat brain, liver and kidney. *Biomed. Pharmacother.* **2017**, *87*, 223–229. [CrossRef]
12. Lin, P.; Chen, F.; Sun, J.; Zhou, J.; Wang, X.; Wang, N.; Li, X.; Zhang, Z.; Wang, A.; Jin, Y. Mycotoxin zearalenone induces apoptosis in mouse Leydig cells via an endoplasmic reticulum stress-dependent signalling pathway. *Reprod. Toxicol.* **2015**, *52*, 71–77. [CrossRef]
13. Kadowaki, H.; Nishitoh, H.; Ichijo, H. Survival and apoptosis signals in ER stress: The role of protein kinases. *J. Chem. Neuroanat.* **2004**, *28*, 93–100. [CrossRef]
14. Malhotra, J.D.; Kaufman, R.J. Endoplasmic reticulum stress and oxidative stress: A vicious cycle or a. double-edged sword? *Antioxid. Redox Signal.* **2007**, *9*, 2277–2293. [CrossRef]
15. Cao, S.S.; Kaufman, R.J. Endoplasmic reticulum stress and oxidative stress in cell fate decision and. human disease. *Antioxid. Redox Signal.* **2014**, *21*, 396–413. [CrossRef]

16. Wang, H.L.; Gao, J.P.; Han, Y.L.; Xu, X.; Wu, R.; Gao, Y. Comparative studies of polydatin and resveratrol on mutual transformation and antioxidative effect *in vivo*. *Phytomedicine* **2015**, *22*, 553–559. [CrossRef] [PubMed]
17. Du, Q.H.; Peng, C.; Zhang, H. Polydatin: A review of pharmacology and pharmacokinetics. *Pharm. Biol.* **2013**, *51*, 1347–1354. [CrossRef]
18. Ji, H.; Zhang, X.; Du, Y.; Liu, H.; Li, S.; Li, L. Polydatin modulates inflammation by decreasing NF-kappaB activation and oxidative stress by increasing Gli1, Ptch1, SOD1 expression and ameliorates blood-brain barrier permeability for its neuroprotective effect in pMCAO rat brain. *Brain Res. Bull.* **2012**, *87*, 50–59. [CrossRef] [PubMed]
19. Jiang, X.; Liu, W.; Deng, J.; Lan, L.; Xue, X.; Zhang, C.; Cai, G.; Luo, X.; Liu, J. Polydatin protects cardiac function against burn injury by inhibiting sarcoplasmic reticulum Ca^{2+} leak by reducing oxidative modification of ryanodine receptors. *Free Radic. Biol. Med.* **2013**, *60*, 292–299. [CrossRef]
20. Ye, J.; Piao, H.; Jiang, J.; Jin, G.; Zheng, M.; Yang, J.; Jin, X.; Sun, T.; Choi, Y.H.; Li, L.; et al. Polydatin inhibits mast cell-mediated allergic inflammation by targeting PI3K/Akt, MAPK, NF-kappaB and Nrf2/HO-1 pathways. *Sci. Rep.* **2017**, *7*, 11895. [CrossRef] [PubMed]
21. Zhao, K.S.; Jin, C.; Huang, X.; Liu, J.; Yan, W.S.; Huang, Q.; Kan, W. The mechanism of Polydatin in shock treatment. *Clin. Hemorheol. Microcirc.* **2003**, *29*, 211–217. [CrossRef] [PubMed]
22. Chen, M.; Hou, Y.; Lin, D. Polydatin Protects Bone Marrow Stem Cells against Oxidative Injury: Involvement of Nrf 2/ARE Pathways. *Stem Cells Int.* **2016**, 9394150. [CrossRef] [PubMed]
23. Bryden, W.L. Mycotoxin contamination of the feed supply chain: Implications for animal productivity and feed security. *Anim. Feed Sci. Technol.* **2012**, *173*, 134–158. [CrossRef]
24. Flores-Flores, M.E.; Lizarraga, E.; de Cerain, A.L.; González-Peñas, E. Presence of mycotoxins in animal milk: A review. *Food Control* **2015**, *53*, 163–176. [CrossRef]
25. Hassen, W.; Ayed-Boussema, I.; Oscoz, A.A.; Ade, C.L.; Bacha, H. The role of oxidative stress in zearalenone-mediated toxicity in Hep G2 cells: Oxidative DNA damage, gluthatione depletion and stress proteins induction. *Toxicology* **2007**, *232*, 294–302. [CrossRef]
26. Wu, M.; Liu, M.; Guo, G.; Zhang, W.; Liu, L. Polydatin Inhibits Formation of Macrophage-Derived Foam Cells. *Evid.-Based Complementary Altern. Med.* **2015**, 729017. [CrossRef]
27. Koneru, M.; Sahu, B.D.; Gudem, S.; Kuncha, M.; Ravuri, H.G.; Kumar, J.M.; Kilari, E.K.; Sistla, R. Polydatin alleviates alcohol-induced acute liver injury in mice: Relevance of matrix metalloproteinases (MMPs) and hepatic anti-oxidants. *Phytomedicine Int. J. Phytother. Phytopharm.* **2017**, *27*, 23–32. [CrossRef]
28. Ince, S.; Avdatek, F.; Demirel, H.H.; Arslan-Acaroz, D.; Goksel, E.; Kucukkurt, I. Ameliorative effect of polydatin on oxidative stress-mediated testicular damage by chronic arsenic exposure in rats. *Andrologia* **2016**, *48*, 518–524. [CrossRef]
29. Liu, H.; Zhao, S.; Zhang, Y.; Wu, J.; Peng, H.; Fan, J.; Liao, J. Reactive oxygen species-mediated endoplasmic reticulum stress and mitochondrial dysfunction contribute to polydatin-induced apoptosis in human nasopharyngeal carcinoma CNE cells. *J. Cell. Biochem.* **2011**, *112*, 3695–3703. [CrossRef] [PubMed]
30. Ben Salah-Abbes, J.; Abbes, S.; Abdel-Wahhab, M.A.; Oueslati, R. Raphanus sativus extract protects against Zearalenone induced reproductive toxicity, oxidative stress and mutagenic alterations in male Balb/c mice. *Toxicon Off. J. Int. Soc. Toxinology* **2009**, *53*, 525–533. [CrossRef]
31. Li, R.P.; Wang, Z.Z.; Sun, M.X.; Hou, X.L.; Sun, Y.; Deng, Z.F.; Xiao, K. Polydatin protects learning and memory impairments in a rat model of vascular dementia. *Phytomedicine Int. J. Phytother. Phytopharm.* **2012**, *19*, 677–681. [CrossRef] [PubMed]
32. Quan, Z.; Gu, J.; Dong, P.; Lu, J.; Wu, X.; Wu, W.; Fei, X.; Li, S.; Wang, Y.; Wang, J.; et al. Reactive oxygen species-mediated endoplasmic reticulum stress and mitochondrial dysfunction contribute to cirsimaritin-induced apoptosis in human gallbladder carcinoma GBC-SD cells. *Cancer Lett.* **2010**, *295*, 252–259. [CrossRef]
33. Ben Salem, I.; Prola, A.; Boussabbeh, M.; Guilbert, A.; Bacha, H.; Abid-Essefi, S.; Lemaire, C. Crocin and Quercetin protect HCT116 and HEK293 cells from Zearalenone-induced apoptosis by reducing endoplasmic reticulum stress. *Cell Stress Chaperones* **2015**, *20*, 927–938. [CrossRef]
34. Banjerdpongchai, R.; Kongtawelert, P.; Khantamat, O.; Srisomsap, C.; Chokchaichamnankit, D.; Subhasitanont, P. Mitochondrial and endoplasmic reticulum stress pathways cooperate in zearalenone-induced apoptosis of human leukemic cells. *J. Hematol. Oncol.* **2010**, *3*, 50. [CrossRef] [PubMed]
35. Zhao, G.; Jiang, K.; Wu, H.; Qiu, C.; Deng, G.; Peng, X. Polydatin reduces Staphylococcus aureus lipoteichoic acid-induced injury by attenuating reactive oxygen species generation and TLR2-NFkappaB signalling. *J. Cell. Mol. Med.* **2017**, *21*, 2796–2808. [CrossRef] [PubMed]
36. Li, J.; Ni, M.; Lee, B.; Barron, E.; Hinton, D.R.; Lee, A.S. The unfolded protein response regulator GRP78/BiP is required for endoplasmic reticulum integrity and stress-induced autophagy in mammalian cells. *Cell Death Differ.* **2008**, *15*, 1460–1471. [CrossRef] [PubMed]
37. Dvash, E.; Har-Tal, M.; Barak, S.; Meir, O.; Rubinstein, M. Leukotriene C4 is the major trigger of stress-induced oxidative DNA damage. *Nat. Commun.* **2015**, *6*, 10112. [CrossRef] [PubMed]

38. Chen, Q.-Q.; Yuan, A.-H.; Yang, J.; Bi-xiang, Z.H.A.; Zhang, M. Effect of acupuncture on the endoplasmic reticulum stress IRE1-CHOP pathway and the expression levels of Bax and Bcl-2 protein as well as genes in pancreatic tissue of rats with diabetes mellitus. *World J. Acupunct. Moxibustion* **2017**, *27*, 41–46. [CrossRef]
39. Liu, Y.H.; Huang, Q.H.; Wu, X.; Wu, J.Z.; Liang, J.L.; Lin, G.S.; Xu, L.Q.; Lai, X.P.; Su, Z.R.; Chen, J.N. Polydatin protects against acetaminophen-induced hepatotoxicity in mice via anti-oxidative and anti-apoptotic activities. *Food Funct.* **2018**, *9*, 5891–5902. [CrossRef]

toxins

Article

Coffee Silverskin and Spent Coffee Suitable as Neuroprotectors against Cell Death by Beauvericin and α-Zearalenol: Evaluating Strategies of Treatment

Ana Juan-García [1,*], Giovanni Caprioli [2], Gianni Sagratini [2], Jordi Mañes [1] and Cristina Juan [1]

[1] Laboratory of Food Chemistry and Toxicology, Faculty of Pharmacy, University of Valencia, Av. Vicent Andrés Estellés s/n, Burjassot, 46100 Valencia, Spain; jordi.manes@uv.es (J.M.); cristina.juan@uv.es (C.J.)
[2] Laboratory of Food Chemistry, School of Pharmacy, University of Camerino, Via S. Agostino 1, 62032 Camerino, Italy; giovanni.caprioli@unicam.it (G.C.); gianni.sagratini@unicam.it (G.S.)
* Correspondence: ana.juan@uv.es

Abstract: Coffee silverskin and spent coffee have been evaluated in a neuroblastoma cell line (SH-SY5Y cells) against beauvericin (BEA) and α-zearalenol (α-ZEL)-induced cytotoxicity with different strategies of treatment. First, the direct treatment of mycotoxins and coffee by-products extracts in SH-SY5Y cells was assayed. IC_{50} values for α-ZEL were 20.8 and 14.0 μM for 48 h and 72 h, respectively and, for BEA only at 72 h, it was 2.5 μM. Afterwards, the pre-treatment with spent coffee obtained by boiling water increased cell viability for α-ZEL at 24 h and 48 h from 10% to 16% and from 25% to 30%, respectively; while with silverskin coffee, a decrease was observed. Opposite effects were observed for BEA where an increase for silverskin coffee was observed at 24 h and 48 h, from 14% to 23% and from 25% to 44%, respectively; however, a decrease below 50% was observed for spent coffee. Finally, the simultaneous treatment strategy for the highest concentration assayed in SH-SY5Y cells provided higher cytoprotection for α-ZEL (from 44% to 56% for 24 h and 48 h, respectively) than BEA (30% for 24 h and 48 h). Considering the high viability of coffee silverskin extracts for SH-SY5Y cells, there is a forthcoming promising use of these unexploited residues in the near future against mycotoxins effects.

Keywords: beauvericin; α-zearalenol; coffee silverskin; spent coffee; SH-SY5Y cells

Key Contribution: SH-SY5Y cells exposed to coffee by-product and mycotoxins in direct treatment, pretreatment and simultaneous treatment allow evaluating the rich polyphenol extracts. Proposal of implementing the Agenda 2030 in SDG#12 and SDG13.

Citation: Juan-García, A.; Caprioli, G.; Sagratini, G.; Mañes, J.; Juan, C. Coffee Silverskin and Spent Coffee Suitable as Neuroprotectors against Cell Death by Beauvericin and α-Zearalenol: Evaluating Strategies of Treatment. *Toxins* **2021**, *13*, 132. https://doi.org/10.3390/toxins13020132

Received: 5 January 2021
Accepted: 5 February 2021
Published: 10 February 2021

Publisher's Note: MDPI stays neutral with regard to jurisdictional claims in published maps and institutional affiliations.

1. Introduction

Coffee is one of the most worldwide consumed beverages. Once the coffee beans are collected and ready to be used, the co-products that remain constitute an underexploited residue that needs to be discarded. Coffee industry and local producers are developing strategies for this large amount of coffee residues that need to be disposed or get used as a valuable nutritional by-product. There two main generated by-products: silverskin coffee and spent coffee grounds. The first one, silverskin, corresponds to the thin tegument that covers the coffee bean. Once green coffee beans are roasted, this subproduct can be obtained; whereas spent coffee grounds can be produced either at home or in the industry during the process to produce instant coffee [1].

The presence of complex amounts of compounds in coffee is extensive, and it is known that different effects are associated; as in gene expression involved in inflammation, immune system, and metabolic pathways [2,3], myocardial blood flow, and associations with liver fibrosis, depression, etc. [4,5]. An understanding of the physiological effects of coffee is drastically limited by the complexities deriving from two factors: the vast array of components included in the brewed product and the varied effects of each compound. Caffeine is the major and active compound in coffee; however, coffee is also rich in other bioactive substances with a wide array of physiological effects. The list comprises up to 1000 described phytochemicals, comprising phenols, including chlorogenic and caffeic acid, lactones, diterpenes, including cafestol and kahweol, niacin, and the vitamin B3 precursor trigonelline [6]. The central nervous system, vascular endothelium, heart, liver, adipose tissue and muscle are tissues containing adenosine receptors, which are the caffeine-target main action [7,8].

Coffee and coffee products can have contaminant fungi and are almost removed by winnowing. The presence of mycotoxins, mainly ochratoxin A or aflatoxins from *Aspergillus* spp., are concentrated in the shell fraction but can reach coffee. When coffee is harvested, prepared, transported, or stored, it is common to have the growth of filamentous fungi as well as in fermentation and drying, especially where the water activity is very low [9]. Thus, industries are focused in having a good mycotoxin HACCP (hazard analysis critical control points) system to prevent the consumer´s exposure as well as to obtain safer products. On the other hand, the protective effect of other substances in food or diet could have an important role. The higher the amount of polyphenols, the higher the capacity to inhibit the effect of mycotoxins and the higher the antioxidant activity.

Mycotoxin effects are very diverse, and some are associated with membrane lipid disturbances with effects on cholesterol-interacting proteins, lipoprotein metabolism, and membrane apo E/amyloid beta interactions relevant to hypercholesterolemia with close connections to neurological diseases [10]. Lipopolysaccharides/mycotoxin interactions interfere with apolipoprotein E/Aβ peptide interactions and determine neuron survival [11]. Mycotoxins pass through the blood–brain barrier and affect astrocytes and oligodendrocytes, whose significant roles are maintenance of the blood–brain barrier integration and nutritive support for the myelin. Consequently, these toxins render myelin susceptible to degradation by various factors [12]. In this sense, the neuroblastoma SH-SY5Y cell line is an accepted cell model to study most cellular alterations linked to neurodegerative diseases and to ameliorate their effects. To notice that once mycotoxins and natural compounds, both present in food, are orally consumed, only those with specific characteristics related to polarity, affinity, pKa value, etc. will reach the brain.

Different *Fusarium* and *Aspergillus* mycotoxins have been involved in exerting toxicity in primary astrocytes [13,14]; some *Penicillium* mycotoxins produce neurological disorders in animals [15], and others have the possibility of passing through the blood–brain barrier [16]. Regarding beauvericin (BEA) and α-zearalenone (α-ZEL), many studies have evaluated their presence and their cytotoxicity in different in vitro cellular lines but not as a possible factor for the neurodegenerative disease and its cell neuronal effect. Most of the assays related to neurotoxicity are based on experimental models; however, for in vitro testing, the neuroblastoma cell line SH-SY5Y is a good biological model to be used as an alternative. It is worth mentioning that the presence of other food components can suppress or enhance their effect on those cells.

In vitro approaches to give a wider use of coffee by-products and with the perspective of their possible neuroprotection from mycotoxins BEA and α-ZEL are here studied. For this purpose, three strategies of treatment: direct treatment, pre-treatment, and simultaneous treatment were followed in a human neuroblasstoma cell line SH-SY5Y exposed to two coffee by-products extracts (silverskin coffee and coffee spent) and two mycotoxins (BEA and α-ZEL). Results obtained within this study are focused on (i) studying the capacity of coffee by-product extract to protect cells from mycotoxin attack evaluating strategies of treatment and (ii) describing the first step to approach mechanisms of action of mycotoxins at the neuronal level as well as coffee by-products extracts.

2. Results

2.1. Cytotoxicity of Mycotoxins and Coffee By-Product Extracts in SH-SY5Y Cells

The cell viability of α-ZEL and BEA is reported in Figure 1. The results clearly indicated that both mycotoxins have toxic effect on SH-SY5Y cells in a dose and time-dependent manner. Figure 1a shows the cytotoxicity of α-ZEL on SH-SY5Y cells and after 48 h and 72 h of exposure; viability went below 50%, reaching IC_{50} values of 20.8 and 14.0 μM, respectively. It can be observed that α-ZEL exerts a stronger cytotoxic effect in SH-SY5Y cells with respect to BEA. In fact, BEA at 2.5 μM and 72 h cell viability effect is maintained above 50% or very close to it. IC_{50} at 72 h was 2.5 μM for BEA (Figure 1b).

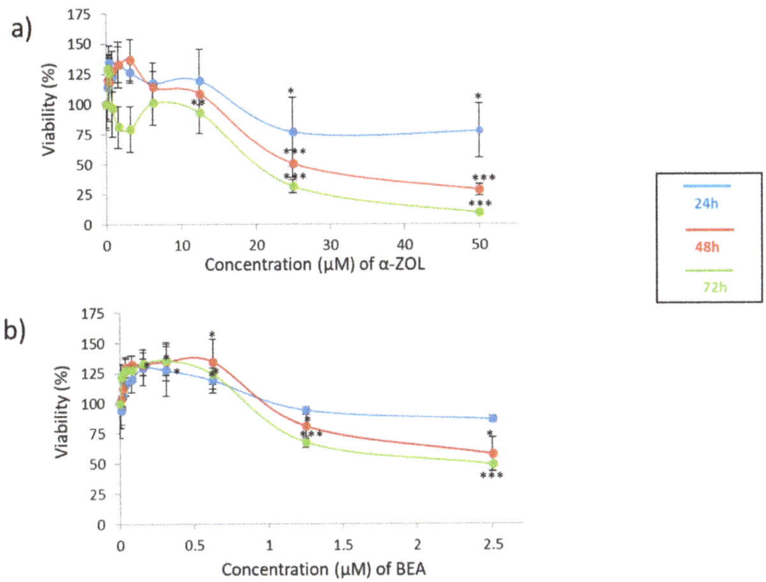

Figure 1. Concentration–effect curves of α-zearalenol (α-ZEL) (**a**) and beauvericin (BEA) (**b**) in SH-SY5Y cells at mycotoxin exposure of 24 h, 48 h and 72 h. The concentration for α-ZEL mycotoxin was 0–50 μM (1:2 dilution), and for BEA, it was 0–2.5 μM (1:2 dilution). * $p \leq 0.05$, ** $p \leq 0.01$ and *** $p \leq 0.001$ represents significant difference as compared to control values.

Coffee extracts were evaluated for their cytotoxicity on SH-SY5Y cells at different dilutions (from 1:0 to 1:16) (data published in Juan et al., 2020). Table 1 reports the order for each pure extract (1:0) of coffee by-products assayed in SH-SY5Y cells according to the viability to clearly show that boiling water provided the higher trend of cell viability in SH-SY5Y cells at 24 h and 48 h. Figure 2 collects the viability of serial dilutions for silverskin and spent coffee extract at 24 and 48 h, evidencing that cell viability in coffee silverskin is higher than in spent coffee extract. At 24 h, as reported in Figure 2a, an increase in cell proliferation of 125% for coffee silverskin extract and 62% for spent coffee extract was observed (dilutions from 1:12 to 1:6); however, from 1:4 to 1:0 dilution, extracts showed a significant reduction in cell proliferation of 25% with respect to the control for spent coffee extract and an increase of 100% with respect to the control for coffee silverskin extract (Figure 2a). At 48 h, an increase in cell proliferation of 50% was detected for both coffee extracts by-products (Figure 2b). Dilutions above 1:8 and 1:4 for coffee silverskin extract and spent coffee extract respectively decreased cell viability. This reduction was highly marked for spent coffee until 20% with respect to the control (Figure 2b). As determined by MTT (3-(4,5-dimethylthiazol-2-yl)-2,5-diphenyltetrazolium bromide) assay only after 48 h, spent coffee extract reached IC_{50} values dilution between 1:2 and 1:0. Controls are referred to cells not treated with mycotoxins, which at the same time did not show differences with the cells exposed to solvent control ($\leq$1%).

Table 1. Viability ranking (order) of coffee by-product pure extracts (at 1:0) exposed to SH-SY5Y cells (boiling water, MeOH, MeOH:H_2O (*v*/*v*, 50:50), EtOH:H_2O (*v*/*v*, 70:30), and MeOH:H_2O (*v*/*v*, 70:30)) at 24 h and 48 h. Gray degradation colors have been assigned for each extract assayed.

	Silverskin Coffee Extracts		Spent Coffee Extract	
Viability Ranking	24 h	48 h	24 h	48 h
1st	Boiling water	Boiling water	Boiling water	EtOH:H_2O (70:30)
2nd	MeOH:H_2O (50:50)	EtOH:H_2O (70:30)	EtOH:H_2O (70:30)	MeOH
3rd	MeOH	MeOH	MeOH:H2O (70:30)	Boiling water
4th	EtOH:H_2O (70:30)	MeOH:H_2O (50:50)	MeOH:H_2O (50:50)	MeOH:H_2O (70:30)
5th	MeOH:H_2O (70:30)	MeOH:H2O (70:30)	MeOH	MeOH:H_2O (50:50)

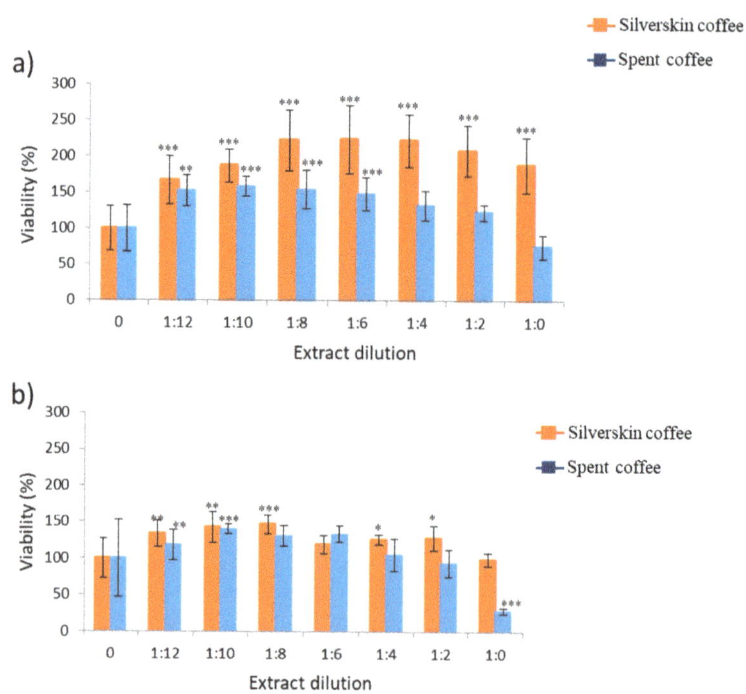

Figure 2. Cytotoxicity of boiling water extract of silverskin (orange bars) coffee and spent coffee (blue bars) on SH-SY5Y cells after 24 h (**a**) and 48 h (**b**) of exposure by (3-(4,5-dimethylthiazol-2-yl)-2,5-diphenyltetrazolium bromide) (MTT) assay. Serial coffee extracts were 1:2 dilutions from 1:0 to 1:12. * $p \leq 0.05$, ** $p \leq 0.01$ and *** $p \leq 0.001$ indicates significant differences compared to the control.

2.2. Cytoprotection Effect of Coffee By-Product Extracts in SH-SY5Y Cells Pre-Treated with α-ZEL and BEA

Boiling water extract was chosen to study cytoprotection, as the direct treatment gave the highest results in cell viability at 1:0 dilution either for 24 and 48 h (Table 1 and Figure 2). Controls refer to cells not treated with mycotoxins, which at the same time did not show differences with the cells exposed to solvent control (≤1%).

2.2.1. Effects of Spent Coffee Boling Water Extract

Figures 3a and 4a show how SH-SY5Y cells pre-treated with spent coffee boiling water extract are affected after exposure to α-ZEL and BEA, respectively. SH-SY5Y cell viability decreased after exposure to α-ZEL and BEA (Figure 1). For α-ZEL, cytoprotection was evidenced in two scenarios where the extract was more concentrated (1:2 and 1:0) and mycotoxins had the highest concentration either for 24 h and 48 h (Figure 3(a.1,a.2)). The marked increase of viability ranged from 10 to 16% and 25 to 30% for 24 h and 48 h, respectively respect to the mycotoxin tested alone.

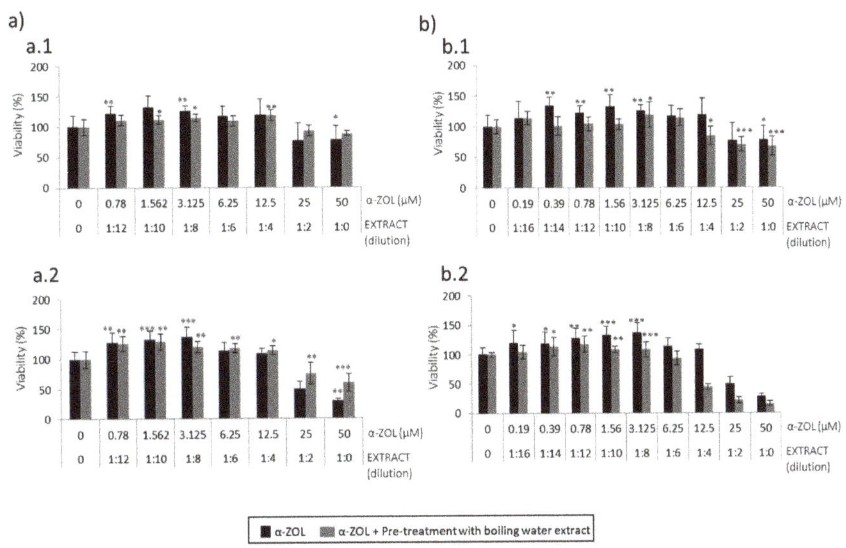

Figure 3. Concentration curves obtained after pre-treatment during 24 h of boiling water extract dilutions of (**a**) spent coffee and (**b**) silverskin coffee, and the subsequent addition of fresh medium with serial dilutions of α-ZEL (starting at 50 μM) during 24 h (**a.1**,**b.1**) and 48 h (**a.2**,**b.2**) in SH-SY5Y cells by MTT assay. All values are expressed as mean ± SD of three replicates (eight wells each time). * $p \leq 0.05$, ** $p \leq 0.01$, and *** $p \leq 0.001$ represent significant difference as compared to control (no treatment).

For BEA, such cytoprotection was not achieved, as no pre-treatment with the boiling water extract of spent coffee reached higher viability values as compared to the mycotoxin tested alone either for 24 h or 48 h (Figure 4a). This might be associated to some type of interaction between spent coffee constituents, BEA mycotoxin, and/or the sensibility of SH-SY5Y. Low viability was detected when SH-SY5Y cells were exposed to BEA 2.5 μM pre-treated with pure coffee extract (1:0): 43% and 14% for 24 and 48 h, respectively (Figure 4(a.1,a.2)).

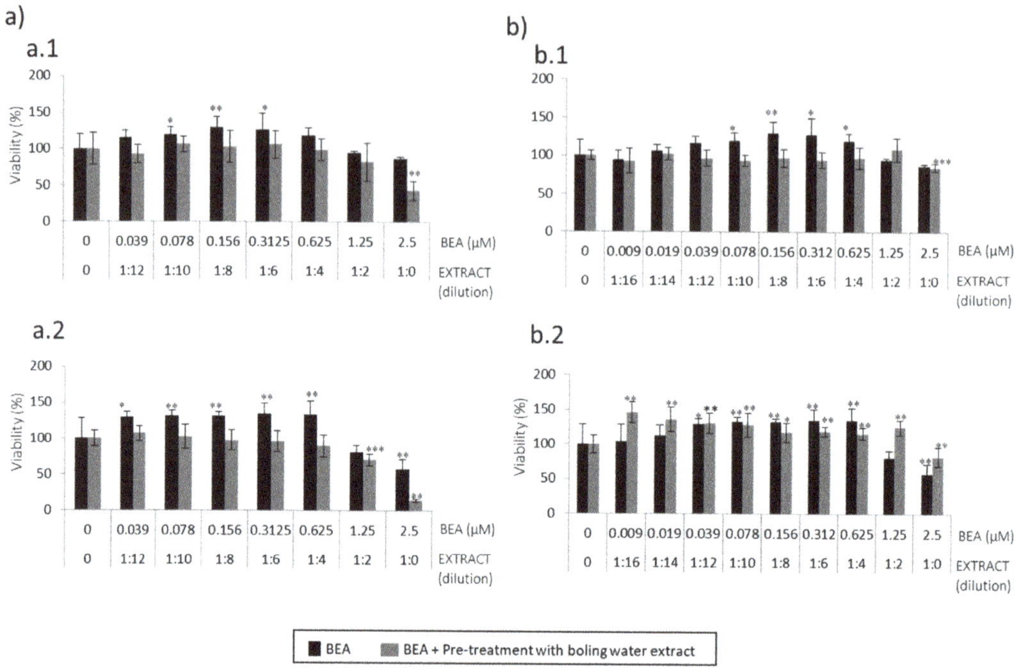

Figure 4. Concentration curves obtained after pre-treatment (24 h) of boiling water extract dilutions (**a**) spent coffee and (**b**) silverskin coffee during 24 h, and subsequent addition of fresh medium with serial dilutions of BEA (starting at 2.5 μM) during 24 h (**a.1**,**b.1**) and 48 h (**a.2**,**b.2**) in SH-SY5Y cells by MTT assay. All values are expressed as mean ± SD of three replicates (eight wells each time). * $p \leq 0.05$, ** $p \leq 0.01$, and *** $p \leq 0.001$ represent significant difference as compared to control (no treatment).

2.2.2. Effects of Coffee Silverskin Boling Water Extract

Figures 3b and 4b show how SH-SY5Y cells pre-treated with coffee silverskin boiling water extract are affected after exposure to α-ZEL and BEA, respectively. For α-ZEL, cytoprotection was not evidenced at 24 h, as viability was maintained very similar in SH-SY5Y cells when pre-treated with coffee extract than with α-ZEL alone (Figure 3(b.1)). Neither did this happen at 48 h, where viability was worst when SH-SY5Y cells were exposed to α-ZEL 50 μM after pre-treatment with pure extract (1:0) viability decreased to 15% (Figure 3(b.2)). It can be hypothesized that there might be some type of interaction with silverskin coffee constituents and α-ZEL mycotoxin.

For BEA, conversely to boiling water of spent coffee extract, cytoprotection was evidenced in the last two scenarios where the extract was more concentrated (1:2 and 1:0) and mycotoxins had the highest concentration either for 24 h and 48 h (Figure 4(b.1,b.2)). The marked increase of viability reached 14% and 44%, for 24 h and 48 h, respectively compared to BEA mycotoxin tested alone.

2.3. *Cytoprotection Effect of Coffee Silverskin Extracts in SH-SY5Y Cells Treated Simultaneously with α-ZEL and BEA*

As the highest viabilities in SH-SY5Y cells were reported for coffee silverskin when boiling water extract was tested, the simultaneous treatment assay was performed with this extract with both mycotoxins (α-ZEL or BEA in Figure 5a,b, respectively) in SH-SY5Y cells. The dilution extract was fixed and tested at 1:4, as its viability is maintained close to 100%. Controls refer to cells not treated with mycotoxins, which at the same time did not show any differences with the cells exposed to solvent control (≤1%).

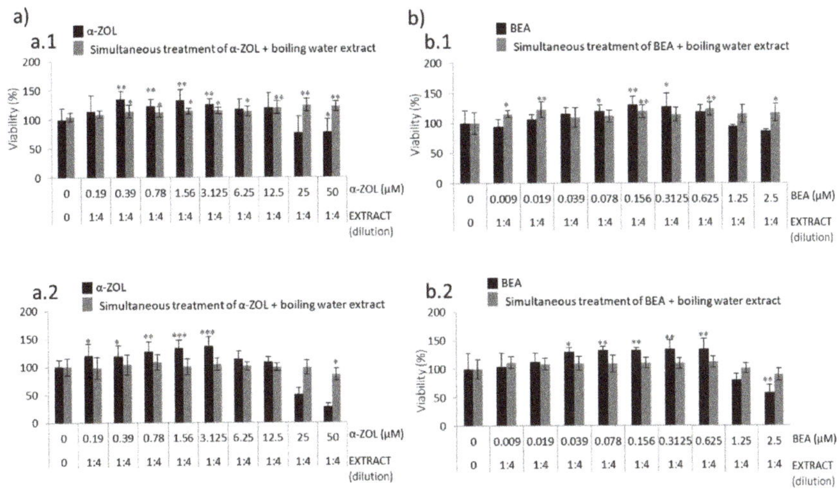

Figure 5. Concentration curves obtained after the simultaneous treatment of boiling water extract (at 1:4 dilution) from silverskin coffee and (**a**) α-ZEL (starting at 50 μM) or (**b**) BEA (starting at 2.5 μM) during 24 h (**a.1**,**b.1**) and 48 h (**a.2**,**b.2**) in SH-SY5Y cells by MTT assay. All values are expressed as mean ± SD of eight replicates. * $p \leq 0.05$, ** $p \leq 0.01$, and *** $p \leq 0.001$ represent significant difference as compared to control (no treatment).

Figure 5a shows the simultaneous treatment of boiling water extract from coffee silverskin and α-ZEL 50 μM 1:2 dilutions at 24 h and 48 h; while Figure 5b shows for BEA 2.5 μM 1:2 dilutions (also for 24 h and 48 h). For α-ZEL and comparing both exposure times, it can be observed that cellular viabilities start to differentiate at 6.25 μM α-ZEL, where there is greater cellular protection (Figure 5a). At both times, when SH-SY5Y cells were exposed to coffee extract 1:4 dilutions and α-ZEL 25 μM and 50 μM simultaneously, the highest protection was reached compared to α-ZEL mycotoxin tested alone as follows: from 48% to 44% and from 40% to 57% for 24 h (Figure 5(a.1)) and 48 h (Figure 5a.2), respectively. For BEA, similar behavior was obtained, and cytoprotection of SH-SY5Y cells was observed for coffee extract 1:4 dilution and BEA at the two highest concentrations tested (Figure 5b). BEA 1.25 μM and 2.5 μM reported a cytoprotection of 20% and 30%, respectively either for 24 h (Figure 5(b.1)) and 48 h (Figure 5b.2) and respect to BEA mycotoxin tested alone. In summary, simultaneous treatment always maintains a higher viability than pre-treatment after 24 h and 48 h for BEA and α-ZEL in SH-SY5Y cells.

Lastly, in the simultaneous study, we compared the behavior of boiling water coffee silverskin extract with EtOH:H_2O (v/v 70:30) and MeOH:H_2O (v/v 40:30) extracts also from coffee silverskin, simulating an identical situation, both at 1:4 fixed extract dilution and α-ZEL (50 μM) mycotoxin and 1:2 dilution (Figure 6).

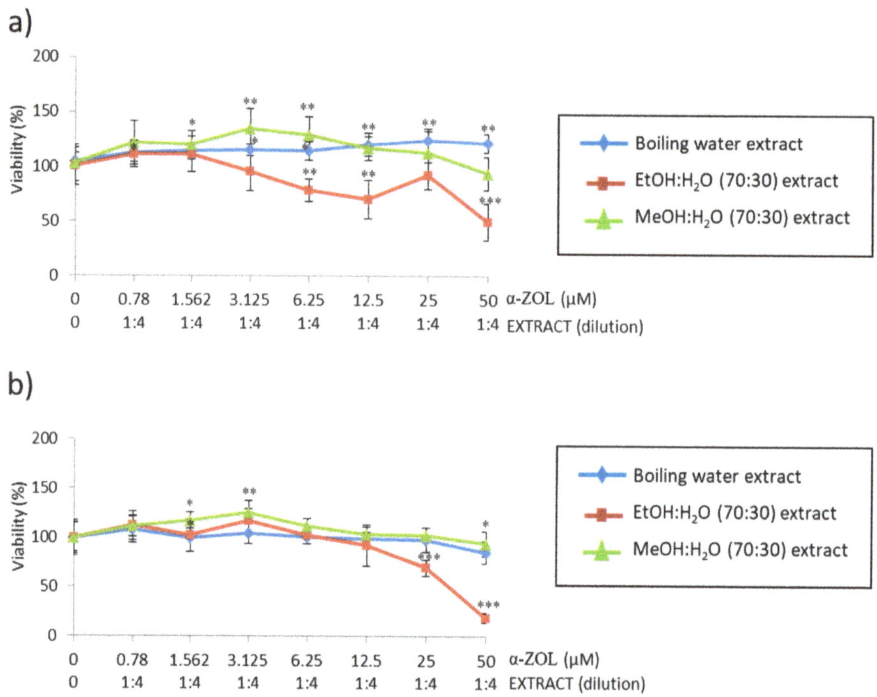

Figure 6. Cytotoxicity of simultaneous treatment of silverskin coffee extracts (boiling water, EtOH:H_2O (70:30) and MeOH:H_2O (70:30)) and α-ZEL (starting at 50 μM) on SH-SY5Y cells after 24 h (**a**) and 48 h (**b**) of exposure by MTT assay. Serial coffee extracts was tested at dilution 1:4. * $p \leq 0.05$, ** $p \leq 0.01$ and *** $p \leq 0.001$ indicates significant differences compared to the control.

Viability of 1:4 dilutions of each extract tested alone in SH-SY5Y cells was above or close to 100% (data not shown) for 24 h and 48 h. At 24 h (Figure 6a) after simultaneous treatment with α-ZEL, the viability of the extracts was maintained very similar from the minimum dilution up to α-ZEL 6.25 μM; from this point, the extract EtOH:H_2O (70:30) is the one that causes a decrease in cellular viability up to a value close to 50% of viability, decreasing at α-ZEL 50 μM until 48%. MeOH:H_2O (70:30) extract initially generates a higher viability up to α-ZEL 12.5μM; then, it decreases the viability to values below those of the boiling water extract (Figure 6a).

At 48 h (Figure 6b), the situation is very similar to that shown at 24 h, because the cellular viability is almost similar in all three extracts up to α-ZEL 12.5 μM; after this point, the viability of the extract boiling water and MeOH:H_2O (70:30) is very similar, while there is a noticeable drop with the extract EtOH:H_2O (70:30), which reaches values of 18%. This means that among all three extracts, at two exposure times, the EtOH:H_2O (70:30) extract is the one that reflects lesser protection against SH-SY5Y cells simultaneously exposed to α-ZEL mycotoxin.

3. Discussion

In the present study, extracts from two coffee by-products (silverskin and spent coffee) were tested by applying different strategies of treatment to evidence whether cytoprotection can be detected in SH-SY5Y mycotoxins treated (α-ZEL and BEA).

There is little literature on mycotoxins in SH-SY5Y cells, and the last report published is for BEA [17]. For zearalenone (ZEA), there is some study on neuronal cells; however, it is the first time that one of its metabolites, α-ZEL, is here studied. The presence of mycotoxins in coffee (grains, brewed, by-products) has been evidenced especially for ochratoxin A, aflatoxins, and fumonisin B2 [18–22]; however, climate change and the importance in recycling and/or reusing residues are redirecting focus to other mycotoxins and the need to study their effect in combination with bioactive compounds.

Despite the lack of studies with mycotoxins in cell lines to report effects of neurotoxicity, the use of alternative models has become crucial. In this sense, the neuroblastoma cell line SH-SY5Y has been used here due to its good characteristics for the exploration of toxicity to humans at this level. Concentrations assayed were chosen according to values of viability reported in MTT cyotoxic assays and close to 80–90% of viability for α-ZEL (see Figure 1a, blue line) and BEA (see Figure 1b, blue line).

The results reported in this study revealed that SH-SY5Y cells respond differently to coffee by-product pure extracts (1:0); nonetheless, the order of higher viability was for most of the scenarios assayed for boiling water (Table 1). In contrast with this, the literature has reported that maximum concentrations of pure coffee silverskin extracts in HepG2 cells were able to reach IC_{50} values [23]. A comparison of two cell lines, Caco-2 and T24 cells, with differences at levels of metabolic capacity, gene expression, differentiation, tumorigenicity, etc. revealed also different sensibility to coffee extracts [24].

Properties associated to coffee grains and brewed coffee have been widely studied, and from the nutritional point of view, the amount of polyphenols as bioactive compounds plays an important role. It has been associated with high amounts of polyphenols with high protecction activity, which results in high cellular proliferation and cytoprotection. This fact was demonstrated for coffee silverskin for most extracts at 1:0 (pure extract) and more specifically for boiling water (Table 1). In a previous study carried out in our lab, total polyphenol content was determined in different extracts. It was revealed that MeOH:H_2O (70:30) extract from silverskin and spent coffee had the highest amount of polyphenols. When this extract from silverskin was assayed in SH-SY5Y cells, as here reported, its viability fell to the last position at 24 h and 48 h with respect to all the other extracts assayed; in spent coffee extract, it was third and fourth position for 24 h and 48 h, respectively, with respect to the other extracts (Table 1). This is also shown here in Figure 2, where silverskin coffee had viability above spent coffee in a great number of scenarios studied. Related to this, assays of green coffee extracts, which are rich in polyphenols, did not increase proliferation in different cell lines such as A549 cells, MRC5 cells, Caco-2 cells, OE-33, T24, and CCD-8Co [24,25]. The main reason associated is the different activities attributed to polyphenols as reported in other studies.

Three strategies of treatment were assayed in SH-SY5Y cells exposed to coffee by-products extract: direct treatment (Figure 2), pre-treatment followed by mycotoxins′ exposure (BEA and α-ZEL) (Figures 3 and 4), and simultaneous treatment with mycotoxins (BEA and α-ZEL) (Figures 5 and 6). The viability of direct treatment of boiling water coffee silverskin extract in SH-SY5Y cells was above spent coffee: from 14% to 100% and from 14% to 70% for 24 h and 48 h, respectively (Figure 2). Similar to this, researchers observed in HeLA cells with spent coffee decreases in a concentration-dependent manner from 40% to 100% at 2 h and 24 h, respectively (Bravo et al., 2013), although extracts of green and robusta coffee in SH-SY5Y cells did not reveal alterations in cell viability [26].

The second strategy carried out was pre-treatment of SH-SY5Y cells with boiling water coffee by-product extracts to be followed by BEA or α-ZEL (Figures 3 and 4) for 24 h. α-ZEL in SH-SY5Y cells previously treated (pre-treatment) with boiling water spent coffee evidenced a relevant cytoprotection without decreasing viability below 50% at 24 h and 48 h, (Figure 3a). This was not found in pre-treatment with coffee silverskin extract, where viability decreased (10%) with respect to the mycotoxin tested alone at 24 h (Figure 3(b.1)). Viability below 50% was reached only at 48 h from 1:4 (12.5 μM) to 1:0 (50 μM) (Figure 3(b.2)). In summary, pre-treatment of SH-SY5Y cells with boiling water coffee silverskin extract and exposed to α-ZEL acts in a less powerful cytoprotection than with spent coffee extract, as lower viability was reached; this was not verified in the same strategy for spent coffee extract (Figure 3a).

For BEA, it was demonstrated that pre-treatment with boiling water of spent coffee extract was much more toxic than with boiling water coffee silverskin extract (Figure 4). At maximum concentrations assayed and after 24 h and 48 h, values below 50% were reached (Figure 4a); while in silverskin, for some of the points assayed (1:0 extract dilution + BEA 2.5 μM and 1:2 extract dilution + BEA 1.25 μM), the extract protected mycotoxins´ effects, which lead to an increase in viability compared to mycotoxin tested alone (Figure 4b). BEA showed a greater capacity to protect SH-SY5Y cells when coffee silverskin extract is tested than when spent coffee extract is tested (Figure 4). Therefore, BEA cytotoxicity in SH-SY5Y cells is alleviated when pre-treatment with boiling water coffee silverskin extract is tested, but not in the case of spent coffee extract. α-ZEL pre-treated with boiling water spent coffee extract was not toxic, but it was when coffee silverskin extract was tested (Figure 3(b.2)).

The last treatment strategy was based on exposing SH-SY5Y cells simultaneously to boiling water coffee by-products extracts of both BEA or α-ZEL (Figure 5). For these assays, the boiling water coffee by-product extract at 1:4 was selected according to the viability reported in Figure 2 (above 50%) and the saturation of the media; 1:4 dilution was maintained fixed for performing the assays simultaneously. In this simultaneous strategy, the treatment of boiling water coffee silverskin extract reached a high cell viability at 24 h as compared to the mycotoxin tested alone in two points (2.5–1.25 μM for BEA and 50–25 μM for α-ZEL) (Figure 5(a.1,b.1)); but after 48 h, only α-ZEL decreased the viability below 50% (Figure 5(a.2)). Comparing the simultaneous treatment of boiling water coffee silverskin extract for both mycotoxins at the highest concentration assayed, it can be observed that simultaneous treatment with α-ZEL (50 μM) exerts greater cellular protection from 44% to 56% with respect to mycotoxin tested alone for 24 h and 48 h (Figure 5a), respectively; while for BEA, the protection was 30% greater with respect to mycotoxin tested alone for both exposure times (Figure 5b).

In a step further in studying the simultaneous strategy of α-ZEL in SH-SY5Y cells, viability was measured and compared with two more coffee silverskin extracts: MeOH:H_2O (70:30) and EtOH: H_2O (70:30) (Figure 6). When comparing these scenarios, boiling water extract kept having the highest viability after 24 h; but after 48 h, it decreased and became higher for MeOH:H_2O (70:30) extract; EtOH:H_2O (70:30) showed the lowest viability both at 24 h and 48 h. Ethanolic extract caused a decrease in cellular viability up to a value slightly lower than the IC_{50} at both exposure times. This means that among all three extracts compared, EtOH:H_2O (70:30) extract does not exert cytoprotection in SH-SY5Y cells following the effect of α-ZEL.

Although there are no previous studies in this regard, investigating mycotoxins exposure to polyphenols and solvent extracts, both pre-treatment and simultaneous treatment had been carried out for evaluating mycotoxins effects. Both have been followed for studying goji berry extracts rich in carotenoids in Caco-2 cells against BEA cytotoxicity [27,28] as well as in HepG2 cells for lentils extracts rich in soyasaponins against alternariol (AOH) cytotoxicity [29]. It reveals that it is a good methodology for evaluating cytoprotection and the potential effects in vitro of bioactive natural compounds.

The effects of antioxidant polyphenolic compounds in cell lines can be found in the literature, but this work is the first to evaluate coffee by-products extracts, silverskin and spent coffee, in SH-SY5Y cells with different strategies of treatments implemented simultaneously and following mycotoxins´ treatment. The results revealed that SH-SY5Y cells respond differently to strategies of treatment and coffee by-product pure extracts (1:0). Nonetheless, it is also evidenced that boiling water extract does not have the entire ability to inhibit the effect of BEA and α-ZEL in all strategies for all scenarios tested, although it was successful for some of them. These facts open the possibility of how to evaluate natural compounds and their potential use from the neuroprotective point of view, and it is starting to captivate the attention of the scientific community. Future perspectives of these studies ought to dive into the capacity of coffee by-products to prevent the formation of reactive oxygen species, while the re-use of these residues contributes to the efficiency of the food industry, being eco-friendly and framed positively within the 2030 Agenda.

4. Materials and Methods

4.1. Chemicals and Reagents

The reagent grade chemicals and cell culture components used were Dulbecco's Modified Eagle's Medium-F12 (DMEM/F-12), fetal bovine serum (FBS), and phosphate buffer saline (PBS), which were supplied by Thermo Fisher, Gibco ™ (Paisley, UK). Methanol (MeOH, HPLC LS/MS grade) was obtained from VWR International (Fontenay-sous-Bois, France). Dimethyl sulfoxide was obtained from Fisher Scientific Co, Fisher BioReagents™ (Geel, Belgium). 3-(4,5-dimethylthiazol- 2-yl)-2,5-diphenyltetrazolium bromide (MTT) for MTT assay, penicillin, streptomycin, and Trypsin–EDTA was purchased from Sigma-Aldrich (St. Louis, MO, USA). Deionized water (<9, MΩcm resistivity) was obtained in the laboratory using a Milli-QSP® Reagent Water System (Millipore, Beadford, MA, USA). The standard of BEA (MW: 783.95 g/mol) and α-ZEL (MW: 320,38 g/mol) were purchased from Sigma-Aldrich (St. Louis, MO, USA). Stock solutions of mycotoxins were prepared in methanol (MeOH) (α-ZEL) and dimethyl sulfoxide (DMSO) (BEA) and maintained at −20 °C in the dark. The final concentration of mycotoxins′ solvents in the medium was ≤1% (*v*/*v*) as per established. All other standards were of standard laboratory grade.

4.2. Cell Culture

Human neuroblastoma cell line, SH-SY5Y, was obtained from American Type Culture Collection (ATCC, Manassas, VA, USA) and cultured in Dulbecco's Modified Eagle's high-glucose medium (DMEM-HAMF-12 4 g/mL), supplemented with 10% fetal bovine serum (FBS) 100 U/mL penicillin, and 100 mg/mL streptomycin. The cells were sub-cultivated after trypsinization once or twice a week and suspended in complete medium in a 1:3 split ratio. Maximum cell passage was 20. Cells were maintained as monolayer in 150 cm^2 cell culture flasks with filter screw caps (TPP, Trasadingen, Switzerland). Cell cultures were incubated at 37 °C, 5% CO_2 atmosphere.

4.3. Coffee By-Product Extracts Rich in Polyphenols

Coffee by-products (coffee silverskin and spent coffee) were received from Camerino (Marche-Umbria Reggio, Italy). Five different solvents and mixtures were used for obtaining extracts from 1 and 10 g of coffee silverskin and spent coffee, respectively. Extracts were prepared by using boiling water, alcohol (MeOH), and hydro-alcoholic mixtures: MeOH:H_2O (*v*/*v*, 70:30), MeOH:H_2O (*v*/*v*, 50:50), and EtOH:H_2O (*v*/*v*, 70:30). Final volumes of 50 mL were collected and concentrated until dryness in a Rotary Evaporator Model R-200 (Büchi, Cornaredo, Italy). The dried coffee extract was dissolved in MeOH and placed into topacium vials, which were previously filtered through a nylon filter 0.22 μm pore size (Analysis Vinicos S.L. Tomelloso, Spain). Extracts were preserved at 4 °C until use.

Extract of coffee by-products were analyzed to determine and identify polyphenols (details of amounts in [30]). These assays were carried out previously in our lab by using Folin–Ciocalteau's reagent [31] and an ultra-high-performance Accurate-Mass Q–TOF-LC/MS analysis (Agilent Technologies, Santa Clara, CA, USA) (see Supplementary Materials Table S1 for details).

4.4. MTT Assay

The MTT assay determines the viability of cells by the reduction of yellow soluble tetrazolium salt (MTT), only in the metabolically active cells, via a mitochondrial-dependent reaction to an insoluble purple formazan crystal. Briefly, after exposure to α-ZEL, BEA, and coffee by-product extracts (from silverskin coffee and spent coffee) by direct treatment, pre-treatment, or simultaneous treatment strategies (described in detail in sections below), the medium containing these compounds was removed and cells of each well received

200 μL fresh medium plus 50 μL of MTT. The plates were wrapped in foil and incubated for 4 h at 37 °C. Afterwards, the medium containing the MTT was removed, and the resulting formazan salt was solubilized in DMSO. The absorbance was measured at 570 nm using an ELISA plate reader Multiscan EX (Thermo Scientific, MA, USA).

4.5. Strategies of Treatment with Beauvericine, α-Zearalenol, and Coffee By-Product Extracts in SH-SY5Y Cells

SH-SY5Y cells were seeded in 96-well culture plates at 2 × 10^4 cells/well and set for 24 h before performing the assays with coffee by-products extracts and mycotoxin´s additions. Subsequently, three different strategies were carried out: direct treatment, pre-treatment, and simultaneous treatment, as explained below.

4.5.1. Strategy of Direct Treatment

Cells seeded in 96-well/plates were treated individually with α-ZEL (from 50 to 0.2 μM) or BEA (from 2.5 to 0.005 μM) at 1:2 serial dilution or coffee extracts (five coffee silverskin extract and five spent coffee extract) starting at the dilution ratio of 1:0 and serial dilutions (until 1:16). When 1:0 dilution was used, the amount of the extract was added to the cell media respecting the proportion of 1% of the entire volume of the well. MTT assay was performed after 24 h and 48 h of exposure. Culture medium without extracts and with 1% solvent were used as control.

4.5.2. Strategy of Pre-Treatment

Pre-treatment consisted of a first step of direct exposure to coffee by-product extracts, as described in previous sub-section. After 24 h, extracts were removed, and the dilution of mycotoxins were added at the same concentration as described previously starting at 50 μM for α-ZEL and at 2.5 μM for BEA 1:2 dilutions all along the entire 96-well plate. MTT assay was performed after 24 h and 48 h. Coffee extracts assayed were both coffee silverskin and spent coffee obtained by a boiling water extraction procedure and subsequent treatment with α-ZEL or BEA. Boiling water extract was selected to follow this strategy, as it gave better viability (protection) and opened the possibility of being used further in the food industry.

4.5.3. Strategy of Simultaneous Treatment

Considering the results obtained in our lab and in Section 2.2.1, dilution used for simultaneous treatment was set at 1:4 for coffee extract by-products, which mantained cell viability above 95%. Dilutions above 1:4 started to decrease cell viability. Then, SH-SY5Y cells were simultaneously treated with boiling water coffee extract (silverskin coffee and spent coffee) at the dilution concentration ratio of 1:4 and mycotoxins at a maximum concentration of 50 μM for α-ZEL and 2.5 μM for BEA and diluted 1:2 for the entire 96-well plate. Controls used were the same as those reported in the "*Direct treatment*" Section. After 24 h and 48 h, the MTT assay was performed. Two more silverskin coffee extracts in simultaneous treatment were compared with boiling water: MeOH:H_2O (70:30, *v*/*v*) and EtOH:H_2O (70:30, *v*/*v*), with α-ZEL mycotoxin. As mentioned before, extracts tested in such treatment were those that gave better viability (protection) or that could report a possibility of using them further in the food industry.

4.6. Statistical Analyses of Data

Statistical analysis of data was carried out using IBM SPSS Statistic version 24.0 (SPSS, Chicago, IL, USA) statistical software package. Data were expressed as mean ± SEM of four independent experiments. The statistical analysis of the results was performed by Student's t-test for paired samples. Differences with respect to the control group were statistically analyzed using ANOVA followed by the Tukey HSD post hoc test for multiple comparisons; $p \leq 0.05$ was considered statistically significant.

Supplementary Materials: The following are available online at https://www.mdpi.com/2072-6651/13/2/132/s1, Table S1: analysis of polyphenols by QTOF –LC/MS.

Author Contributions: Data curation: A.J.-G. and C.J.; Formal analysis: A.J.-G. and C.J.; Funding acquisition: G.C. and J.M.; Methodology: A.J.-G. and C.J.; Project administration: G.C. and J.M.; Resources: J.M.; Supervision: C.J., G.S. and J.M.; Visualization: G.C., G.S. and C.J.; Writing-original draft: A.J.-G. and C.J.; writing-review & editing: A.J.-G., C.J., G.S. and J.M. All authors have read and agreed to the published version of the manuscript.

Funding: This work has been funded by the Spanish Ministry of Economy and Competitiveness PID2019-108070RB-100AL and University of Camerino (Fondo di Ateneo per la Ricerca—Year 2018, Grant no. FPI000051).

Institutional Review Board Statement: Not applicable.

Informed Consent Statement: Not applicable.

Data Availability Statement: Not applicable.

Acknowledgments: The authors would like to thank Spanish Ministry of Economy and Competitiveness for the funding received to perform this study and Fondo di Ateneo per la Ricerca from the University of Camerino.

Conflicts of Interest: The authors declare no conflict of interest. Compliance with ethical standards.

References

1. Galanakis, C.M. *Handbook of Coffee Processing by-Products: Sustainable Applications*; Academic Press Elsevier: Cambridge, MA, USA, 2017.
2. Barnung, R.; Nøst, T.; Ulven, S.M.; Skeie, G.; Olsen, K. Coffee consumption and whole-blood gene expression in the norwegian women and cancer post-genome cohort. *Nutrients* **2018**, *10*, 1047. [CrossRef]
3. Kuang, A.; Erlund, I.; Herder, C.; Westerhuis, J.A.; Tuomilehto, J.; Cornelis, M.C. Lipidomic response to coffee consumption. *Nutrients* **2018**, *10*, 1851. [CrossRef] [PubMed]
4. Van Dijk, R.; Ties, D.; Kuijpers, D.; Van der Harst, P.; Oudkerk, M. Effects of caffeine on myocardial blood flow: A systematic review. *Nutrients* **2018**, *10*, 1083. [CrossRef] [PubMed]
5. Cornelis, M.C. The impact of caffeine and coffee on human health. *Nutrients* **2019**, *13*, 416. [CrossRef] [PubMed]
6. Cano-Marquina, A.; Tarín, J.J.; Cano, A. The impact of coffee on health. *Maturitas* **2013**, *75*, 7–21. [CrossRef]
7. Yaya, I.; Marcellin, F.; Costa, M.; Morlat, P.; Protopopescu, C.; Pialoux, G.; Santos, M.E.; Wittkop, L.; Esterle, L.; Gervais, A. Impact of alcohol and coffee intake on the risk of advanced liver fibrosis: A longitudinal analysis in hiv-hcv coinfected patients (anrs hepavih co-13 cohort). *Nutrients* **2018**, *10*, 705. [CrossRef] [PubMed]
8. Haller, S.; Montandon, M.L.; Rodriguez, C.; Herrmann, F.R.; Giannakopoulos, P. Impact of coffee, wine, and chocolate consumption on cognitive outcome and MRI parameters in old age. *Nutrients* **2018**, *10*, 1391. [CrossRef]
9. Bessaire, T.; Perrin, I.; Tarres, A.; Bebius, A.; Reding, F.; Theurilla, V. Mycotoxins in green coffee: Occurrence and risk assessment. *Food Control* **2019**, *96*, 59–67. [CrossRef]
10. Cheng, J.; Huang, S.; Fan, C.; Zheng, N.; Zhang, N.; Li, S.; Wang, J. Metabolomic analysis of alterations in lipid oxidation, carbohydrate and amino acid metabolism in dairy goats caused by exposure to aflatoxin B1. *J. Dairy Res.* **2017**, *84*, 401–406. [CrossRef]
11. Martins, I.J. Overnutrition Determines LPS Regulation of Mycotoxin Induced Neurotoxicity in Neurodegenerative Diseases. *Int. J. Mol. Sci.* **2015**, *16*, 29554–29573. [CrossRef]
12. Purzycki, C.B.; Shain, D.H. Fungal toxins and multiple sclerosis: A compelling connection. *Brain Res. Bull.* **2010**, *82*, 4–6. [CrossRef] [PubMed]
13. Stockmann-Juvala, H.; Savolainen, K. A review of the toxic effects and mechanisms of action of fumonisin B1. *Hum. Exp. Toxicol.* **2008**, *27*, 799–809. [CrossRef] [PubMed]
14. Weidner, M.; Lenczyk, M.; Schwerdt, G.; Gekle, M.; Humpf, H.-U. Neurotoxic potential and cellular uptake of T-2 toxin in human astrocytes in primary culture. *Chem. Res. Toxicol.* **2013**, *26*, 347–355. [CrossRef] [PubMed]
15. Cavanagh, J.B.; Holton, J.L.; Nolan, C.C.; Ray, D.E.; Naik, J.T.; Mantle, P.G. The effects of the tremorgenicmycotoxin penitrem A on the rat cerebellum. *Vet. Pathol.* **1998**, *35*, 53–63. [CrossRef]
16. Behrens, M.; Hüwel, S.; Galla, H.-J.; Humpf, H.U. Blood-brain barrier effects of the Fusarium mycotoxins deoxynivalenol, 3-acetyldeoxynivalenol, and moniliformin and their transfer to the brain. *PLoS ONE* **2015**, *10*, e0143640. [CrossRef]
17. Agahi, F.; Font, G.; Juan, C.; Juan-García, A. Individual and combined effect of zearalenone derivates and beauvericin mycotoxins on SH-SY5Y cells. *Toxins* **2020**, *12*, 212. [CrossRef]
18. García-Moraleja, A.; Font, G.; Manes, J.; Ferrer, E. Simultaneous determination of mycotoxin in commercial coffee. *Food Control* **2015**, *57*, 282–292. [CrossRef]

19. Juan, C.; Covarelli, L.; Beccari, G.; Colasante, V.; Mañes, J. Simultaneous analysis of twenty-six mycotoxins in durum wheat grain from Italy. *Food Control* **2016**, *62*, 322–329. [CrossRef]
20. Juan, C.; Berrada, H.; Mañes, J.; Oueslati, S. Multi-mycotoxin determination in barley and derived products from Tunisia and estimation of their dietary intake. *Food Chem. Toxicol.* **2017**, *103*, 148–156. [CrossRef]
21. Oueslati, S.; Berrada, H.; Mañes, J.; Juan, C. Presence of mycotoxins in Tunisian infant foods samples and subsequent risk assessment. *Food Control* **2017**, *84*, 362–369. [CrossRef]
22. Stanciu, O.; Juan, C.; Miere, D.; Loghin, F.; Mañes, J. Presence of enniatins and beauvericin in Romanian wheat samples: From raw material to products for direct human consumption. *Toxins* **2017**, *9*, 189. [CrossRef] [PubMed]
23. Iriondo-DeHond, A.; Haza, A.I.; Ávalo, A.; Del Castillo, M.D.; Morales, P. Validation of coffee silverskin extract as a food ingredient by the analysis of cytotoxicity and genotoxicity. *Food Res. Int.* **2017**, *100*, 791–797. [CrossRef] [PubMed]
24. Amigo-Benavent, M.; Wang, S.; Mateos, R.; Sarrià, B.; Bravo, L. Antiproliferative and cytotoxic effects of green coffee and yerba mate extracts, their main hydroxycinnamic acids, methylxanthine and metabolites in different human cell lines. *Food Chem. Toxicol.* **2017**, *106*, 125–138. [CrossRef]
25. Burgos-Morón, E.; Calderón-Montaño, J.M.; Orta, M.L.; Pastor, N.; Pérez-Guerrero, C.; Austin, C.; Mateos, S.; López-Lázaro, M. The coffee constituent chlorogenic acid induces cellular DNA damage and formation of topoisomerase I− and II−DNA complexes in cells. *J. Agric. Food Chem.* **2012**, *60*, 7384–7391. [CrossRef]
26. Ciaramelli, C.; Palmioli, A.; De Luigi, A.; Colombo, L.; Sala, G.; Riva, C.; Zoia, C.P.; Salmona, M.; Airoldi, C. NMR-driven identification of anti-amyloidogenic compounds in green and roasted coffee extracts. *Food Chem.* **2018**, *252*, 171–180. [CrossRef]
27. Montesano, D.; Juan-García, A.; Mañes, J.; Juan, C. Chemoprotective effect of carotenoids from *Lycium barbarum* L. on SH-SY5Y neuroblastoma cells treated with beauvericin. *Food Chem. Toxicol.* **2020**, *141*, 111414. [CrossRef] [PubMed]
28. Juan-García, A.; Montesano, D.; Mañes, J.; Juan, C. Cytoprotective effects of carotenoids-rich extract from *Lycium barbarum* L. on the beauvericin-induced cytotoxicity on Caco-2 cells. *Food Chem. Toxicol.* **2019**, *133*, 110798. [CrossRef] [PubMed]
29. Vila-Donat, P.; Fernandez-Blanco, C.; Sagratini, G.; Font, G.; Ruiz, M.J. Effects of soyasaponin I and soyasaponins-rich extract on the Alternariol-induced cytotoxicity on Caco-2 cells. *Food Chem. Toxicol.* **2015**, *77*, 44–49. [CrossRef]
30. Juan, C.; De Simone, G.; Sagratini, G.; Caprioli, G.; Manes, J.; Juan-García, A. Reducing the effect of beauvericin on neuroblastoma SH-SY5Y cell line by natural products. *Toxicon* **2020**, *188*, 164–171. [CrossRef]
31. Giusti, F.; Caprioli, G.; Ricciutelli, M.; Vittori, S.; Sagratini, G. Determination of fourteen polyphenols in pulses by high performance liquid chromatography-diode array detection (HPLC-DAD) and correlation study with antioxidant activity and colour. *Food Chem.* **2017**, *221*, 689–697. [CrossRef]

toxins

MDPI

Article

Zearalenone Induces Endothelial Cell Apoptosis through Activation of a Cytosolic Ca^{2+}/ERK1/2/p53/Caspase 3 Signaling Pathway

Hyeon-Ju Lee, Se-Young Oh and Inho Jo *

Graduate Program in System Health Science and Engineering, Department of Molecular Medicine, College of Medicine, Ewha Womans University, 25 Magokdong-ro-2-gil, Gangseo-gu, Seoul 07804, Korea; eihj0323@hanmail.net (H.-J.L.); ohs@ewha.ac.kr (S.-Y.O.)

* Correspondence: inhojo@ewha.ac.kr

Abstract: Zearalenone (ZEN) is a mycotoxin that has been reported to damage various types of cells/tissues, yet its effects on endothelial cells (ECs) have never been investigated. Therefore, this study investigates the potential effects of ZEN using bovine aortic ECs (BAECs). In this study, we found that ZEN induced apoptosis of BAECs through increased cleavage of caspase 3 and poly ADP-ribose polymerase (PARP). ZEN also increased phosphorylation of ERK1/2 and p53, and treatment with the ERK1/2 or p53 inhibitor reversed ZEN-induced EC apoptosis. Transfection of BAECs with small interfering RNA against ERK1/2 or p53 revealed ERK1/2 as an upstream target of p53 in ZEN-stimulated apoptosis. ZEN increased the production of reactive oxygen species (ROS), yet treatment with the antioxidant did not prevent EC apoptosis. Similarly, blocking of estrogen receptors by specific inhibitors also did not prevent ZEN-induced apoptosis. Finally, chelation of cytosolic calcium (Ca^{2+}) using BAPTA-AM or inhibition of endoplasmic reticulum (ER) Ca^{2+} channel using 2-APB reversed ZEN-induced EC apoptosis, but not by inhibiting ER stress using 4-PBA. Together, our findings demonstrate that ZEN induces EC apoptosis through an ERK1/2/p53/caspase 3 signaling pathway activated by Ca^{2+} release from the ER, and this pathway is independent of ROS production and estrogen receptor activation.

Keywords: mycotoxin; zearalenone; apoptosis; endothelial cells; calcium

Key Contribution: (1) ZEN induced-EC apoptosis is independent of ROS production and estrogen receptor. (2) ZEN activates ER Ca^{2+} channel to increase cytosolic Ca^{2+}, triggering apoptosis. (3) ZEN-induced apoptosis of BAECs in a sequence of cytosolic Ca^{2+}/p53/ERK/caspase 3.

Citation: Lee, H.-J.; Oh, S.-Y.; Jo, I. Zearalenone Induces Endothelial Cell Apoptosis through Activation of a Cytosolic Ca^{2+}/ERK1/2/p53/Caspase 3 Signaling Pathway. *Toxins* **2021**, *13*, 187. https://doi.org/10.3390/toxins13030187

Received: 14 January 2021
Accepted: 28 February 2021
Published: 4 March 2021

1. Introduction

Zearalenone (ZEN) is a common mycotoxin produced by the *Fusarium* species, and is a frequent contaminant of crops, grains and food products [1]. ZEN can be converted into reduced metabolites, such as α-zearalenol, β-zearalenol, α-zearalanol, β-zearalanol and zearalanone [2]. ZEN and some of its metabolites are known to exhibit estrogenic activity and disrupt endocrine functions in animals [3,4]. The molecular mode of action of ZEN in mimicking estrogen activity has been established by many studies. It has been proposed that ZEN binds competitively to genomic or nongenomic estrogen receptors to activate transcription of estrogen-responsive genes, eventually leading to estrogen/estrogen receptor-mediated cellular effects [5,6]. These actions may consequently disrupt reproductive functions in variety of cells and animals; for example, intake of ZEN for 70 days decreased sperm concentration and altered the morphology of spermatozoa in male mice [7].

Apart from the estrogenic activity of ZEN, ZEN also mediates other toxicological effects through various mechanisms [8,9]. For example, ZEN increases intracellular reactive oxygen species (ROS) production, in various cell types, including porcine small intestinal epithelial cell line SIEC02, mouse Sertoli TM4 cells [10] and human neuroblastoma cell line SHSY-5Y cells [11]. ZEN-induced ROS production also mediates endoplasmic reticulum (ER) stress to reduce proliferation and increase apoptosis of mouse Leydig cells exposed to ZEN in vitro [12]. Another study also showed that ZEN exposure induces pro-apoptotic protein Bax, promoting cytochrome c release from mitochondria into the cytosol, which activates caspase 9 and 3 followed by cleavage of poly ADP-ribose polymerase (PARP) in Leydig cells [13]. Other apoptotic transcription factors, such as the c-jun NH_2-terminal kinase (JNK) and p38 mitogen-activated protein kinase (MAPK) (p38 MAPK), were also shown to be induced in murine macrophage cell line RAW264.7 exposed to ZEN [14]. These studies clearly suggest that ZEN activates various molecular pathways to induce cell apoptosis, thereby damaging its target cells/tissues.

Many studies have used endocrine and reproductive cells to assess the toxicological effects of ZEN, but the potential effects on endothelial cells (ECs) have been neglected, despite their potential biological relevance. ECs are one of essential components of the vascular system. Once drugs and/or chemicals are ingested, they first contact the EC lining of the blood vessels and other cellular matrices before reaching their target sites. Studies have shown that ECs can be easily damaged by various stressors, such as cigarette aerosol, H_2O_2, and oxygen-glucose deprivation/reperfusion [15–17]. These factors can cause DNA damage and inflammation, which eventually lead to death of ECs, and weaken the integrity and function of blood vessels.

ZEN is often found in animal feed, altering the reproductive function of animals. Previous studies showed that ZEN changes the serum level of progesterone and estradiol and induces teratogenic effects in pigs and sheep [18]. Bovine species are also considered as a major exposure group to mycotoxins including ZEN, and bovine aortic ECs (BAECs) have been suggested as a good model for functional studies on ECs due to their ease of using molecular techniques such as gene transfection and modification. For these reasons, we investigated the potential effects of ZEN using BAECs.

2. Results

2.1. Zearalenone Increases Apoptotic Death of BAECs

ZEN has diverse effects depending on species, cell types as well as exposure concentrations and time. Concentration ranges of ZEN between 10 pM and 300 μM have generally been used to assess acute toxicity of ZEN in several previous studies [19]. Our study tested various exposure conditions of ZEN on BAECs, and determined that ZEN reduced the viability of BAECs in a concentration- and time-dependent manner (Figure 1a,b). Treatment with 30 μM ZEN for 24 h significantly reduced BAEC viability to ~60% compared with the unexposed control group, and thus, all subsequent experiments were carried out using 30 μM ZEN treatment for 24 h, unless specifically stated otherwise. As shown in Figure 1c,d, ZEN increased cleavage of caspase 3 and PARP in a concentration- and time-dependent manner. An apoptosis assay using annexin V-FITC/PI staining revealed that 2.22, 7.31, 21.82 and 24.68% of BAECs underwent apoptotic cell death when treated with 0, 10, 30 and 60 μM of ZEN, respectively (Figure 1e). When Z-DEVD-FMK, a caspase 3 inhibitor, was co-treated with 30 μM ZEN, cleavage of caspase 3 and PARP was significantly prevented (Figure 1f), and apoptotic cell death induced by ZEN was reduced from 31.62% to 5.43% (Figure 1g). These results indicate that ZEN decreases BAEC viability at least partially through a caspase 3-dependent apoptotic pathway.

2.2. ZEN-Induced Apoptosis Is Mediated through Phosphorylation of ERK1/2

The MAPK proteins JNK, p38 MAPK and ERK1/2 have been reported to be common apoptotic signaling molecules [20,21]. In particular, JNK and p38 MAPK were reported to mediate ZEN-induced apoptosis in RAW264.7 cells [14]. Based on these previous findings, we examined the role of JNK and p38 MAPK in ZEN-induced apoptosis of the BAECs. However, neither treatment with SP600125, a JNK inhibitor, nor SB203580, a p38 MAPK inhibitor, inhibited the effect of ZEN on cleavage of caspase 3 and PARP, suggesting no involvement of these two kinases in the ZEN-induced apoptosis (Figure 2a). When ERK1/2, another protein of the MAPK family, was selectively inhibited using U0126, the expression of cleaved caspase 3 and PARP induced by ZEN exposure was reduced (Figure 2b). As expected, treatment with U0126 also prevented the apoptosis induced by ZEN treatment (Figure 2c). Together, these data suggested that ERK1/2 plays an important role in enhancing ZEN-induced apoptosis of BAECs.

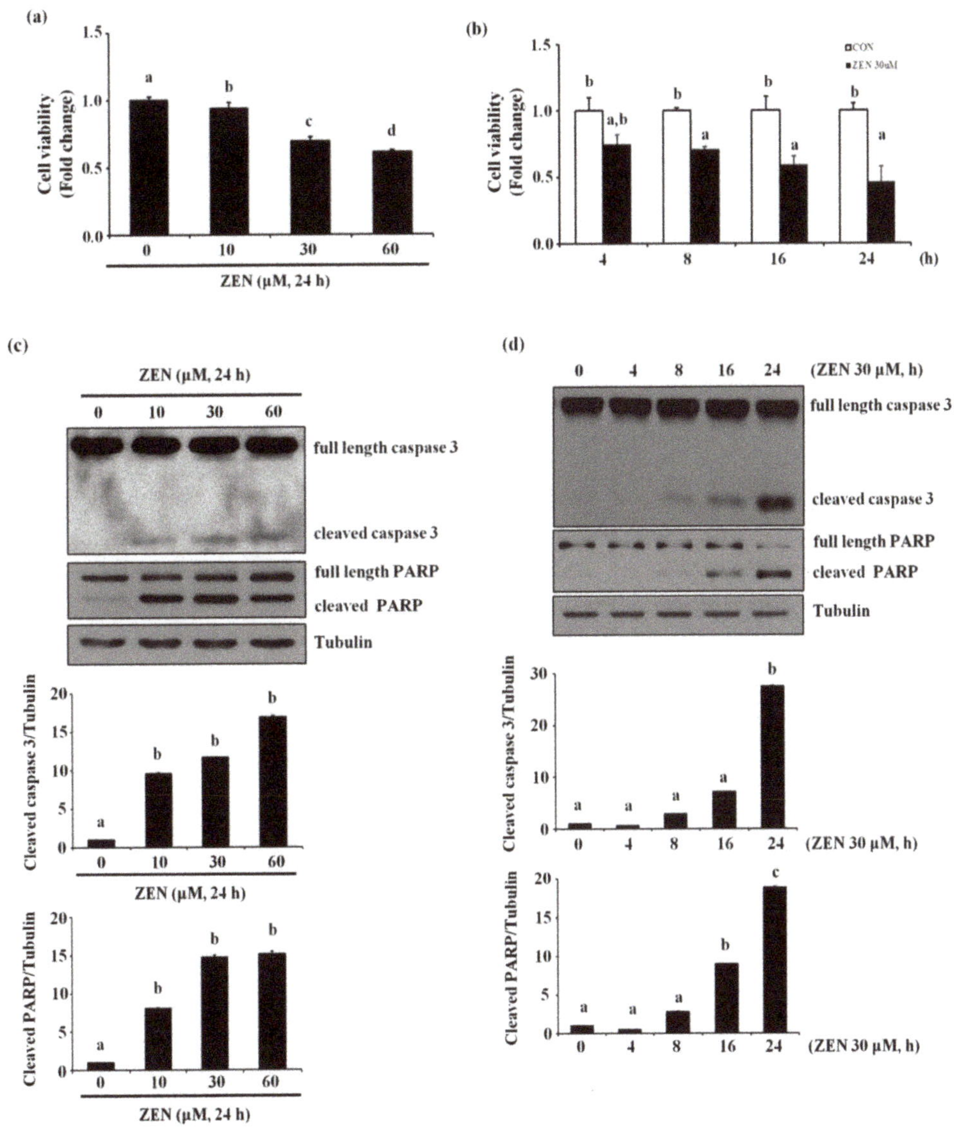

Figure 1. *Cont.*

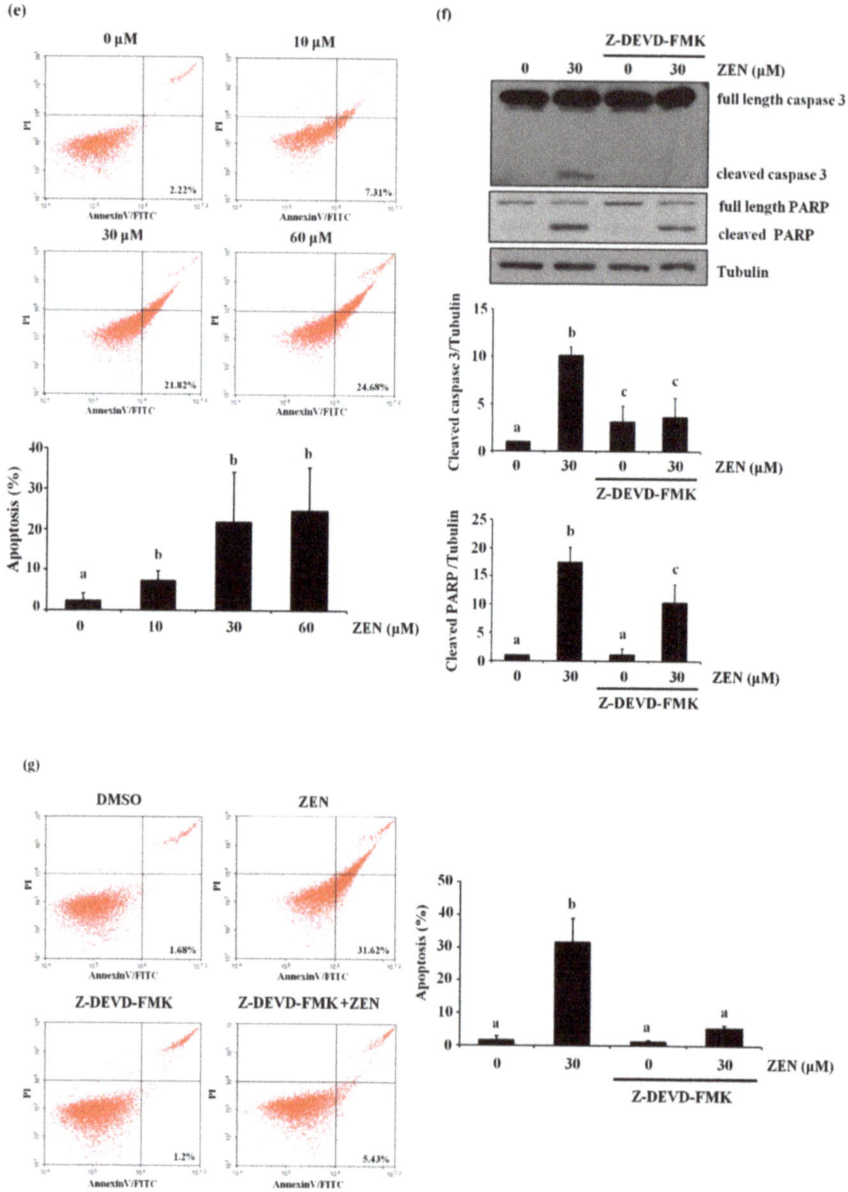

Figure 1. ZEN reduces the viability of BAECs by caspase-dependent apoptosis. BAECs were treated with various concentrations of ZEN (0, 10, 30 or 60 μM) for 24 h or 30 μM of ZEN for various time points (0, 4, 8, 16 or 24 h). (**a**,**b**) Cell viability was measured using the MTT assay. (**c**,**d**) The protein expression of cleaved caspase 3 and PARP in the BAECs was quantified (relative to relative to tubulin) using western blot analyses. (**e**) Apoptosis induced by ZEN at different concentrations was measured by FACS using annexin V/PI staining. After the pretreatment of 20 μM Z-DEVD-FMK for 1 h, BAECs were incubated with 30 μM ZEN for 24 h. (**f**) The protein expression of cleaved caspase 3 and PARP relative to tubulin was quantified using western blot analyses. (**g**) Apoptosis was measured by FACS using annexin V/PI staining. The plots depict the mean fold changes relative to the control (±SD) from at least four independent experimental trials. The different alphabetical letters refer to significant difference ($p < 0.05$) among groups, which were determined by one-way ANOVA followed by Tukey's multiple comparisons.

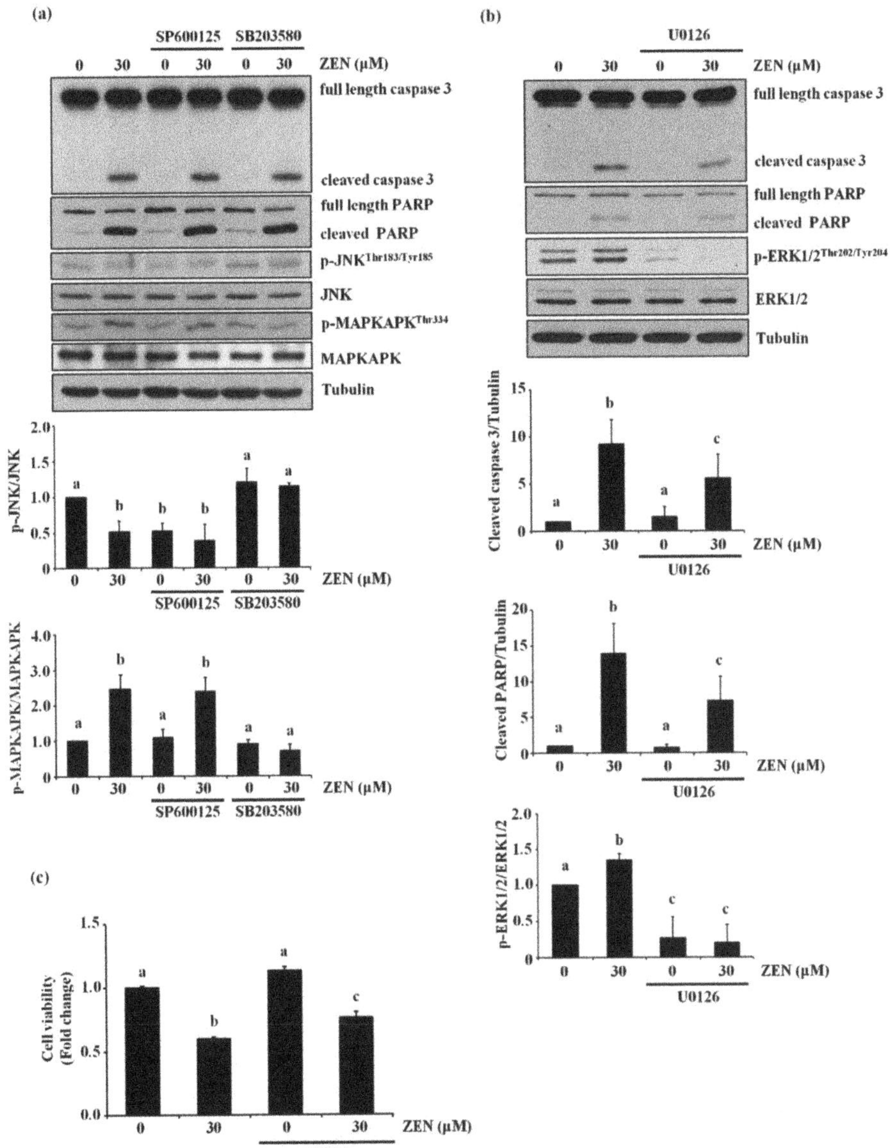

Figure 2. ERK1/2 mediates apoptosis by ZEN in BAECs. **(a)** The protein expression of cleaved caspase 3 and PARP relative to tubulin was quantified using western blot analyses in BAECs exposed to 30 μM of ZEN for 24 h after pretreatment with JNK inhibitor SP600125 (1 μM) or p38 MAPK inhibitor SB203580 (5 μM) for 1 h. SB203580 does not directly affect phosphorylation of p38, but inhibits p38 catalytic activity by binding to the ATP binding pocket, inhibiting phosphorylation of MAPKAPK, a downstream molecule of p38 MAPK. The plots are representative of at least four independent experimental trials. **(b)** After pretreatment with 1 μM of ERK1/2 inhibitor U0126 for 1 h, BAECs were incubated with 30 μM of ZEN for 24 h. The protein expression of cleaved caspase 3 and PARP, p-ERK1/2, and ERK1/2 relative to tubulin was quantified using western blot analyses. **(c)** Cell viability was measured using the MTT assay. The plots depict the mean fold changes relative to the control (±SD) from at least four independent experimental trials. The different alphabetical letters refer to significant difference ($p < 0.05$) among groups, which were determined by one-way ANOVA followed by Tukey's multiple comparisons.

2.3. p53 Is Involved in ZEN-Induced Apoptosis of BAECs

Since p53 is reported to mediate ZEN-induced apoptosis in HepG2 and RAW264.7 cells [14,22], we examined whether p53 was also involved in ZEN-induced EC apoptosis under our experimental conditions. Inhibition of p53 using pifithrin-α reversed the apoptosis induced by ZEN as shown in Figure 3a. As expected, we also found that cleavage of caspase 3 and PARP was significantly reduced in cells treated with the p53 inhibitor (Figure 3b). These results indicated that p53 was involved in ZEN-induced apoptosis of BAECs.

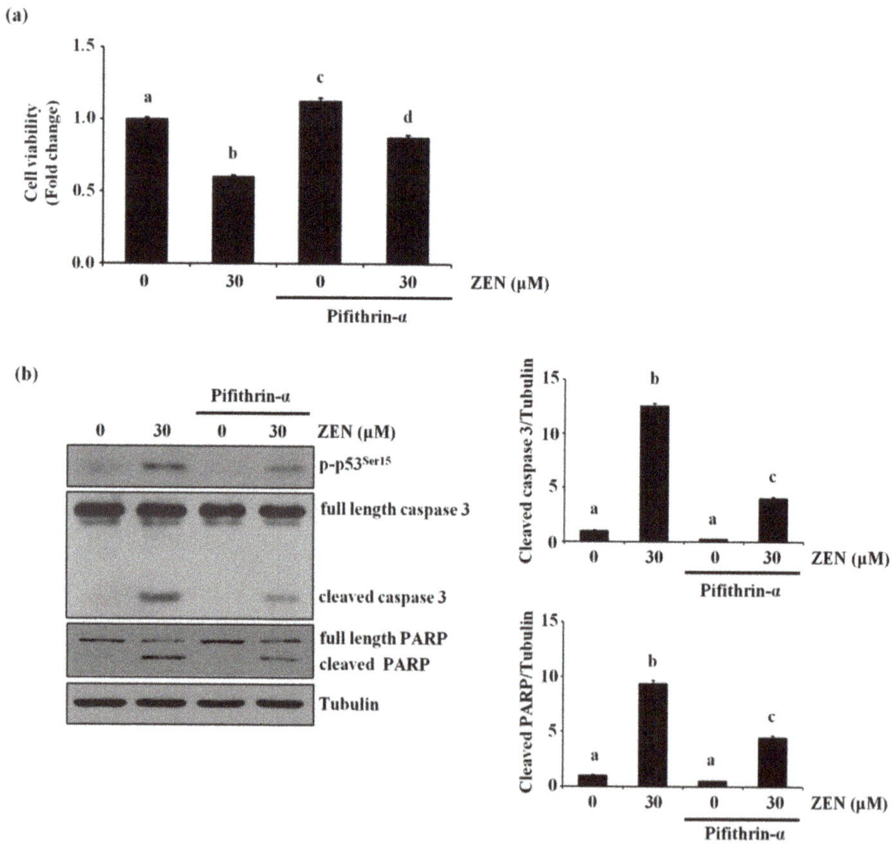

Figure 3. p53 is involved in EC apoptosis by ZEN. After pretreatment with 5 μM of p53 inhibitor pifithrin-α for 1 h, BAECs were incubated with 30 μM of ZEN for 24 h. (**a**) Cell viability was measured by using the MTT assay. (**b**) The protein expression of cleaved caspase 3 and PARP, and p-p53 relative to tubulin was quantified using western blot analyses. The plots depict the mean fold changes relative to the control (±SD) from at least four independent experimental trials. The different alphabetical letters refer to significant difference ($p < 0.05$) among groups, which were determined by one-way ANOVA followed by Tukey's multiple comparisons.

2.4. ERK1/2 Is an Upstream Mediator of p53-Mediating ZEN-Induced EC Apoptosis

In light of the above results showing the involvement of two proteins, ERK1/2 and p53, in the ZEN-induced apoptosis, we explored which of the two molecules was the upstream mediator of this signaling pathway. As shown in Figure 4a, the inhibition of ERK1/2 using U0126 prevented ZEN-induced phosphorylation of p53. In contrast, inhibition of p53 using pifithrin-α did not alter ZEN-induced phosphorylation of ERK1/2. These results suggest that phosphorylation of p53 is partially dependent on the phosphorylation of ERK1/2 in the ZEN-induced apoptotic pathway. This finding was further validated by the use of siRNA against ERK1/2 or p53 (Figure 4b–d), where siRNA against ERK1/2 reduced ZEN-induced phosphorylation of p53 but not vice versa. Our results indicated that ZEN induces apoptotic death of BAECs by a signaling axis of the ERK1/2/p53/caspase 3.

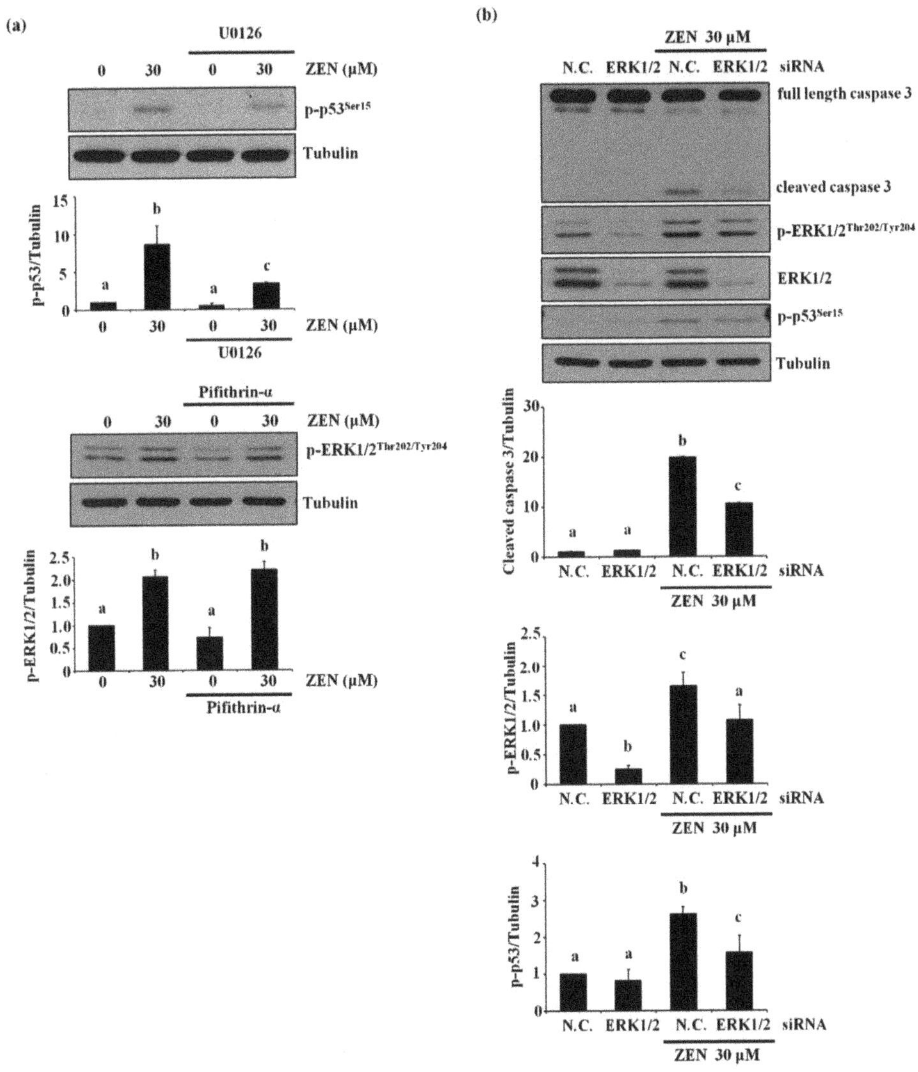

Figure 4. *Cont.*

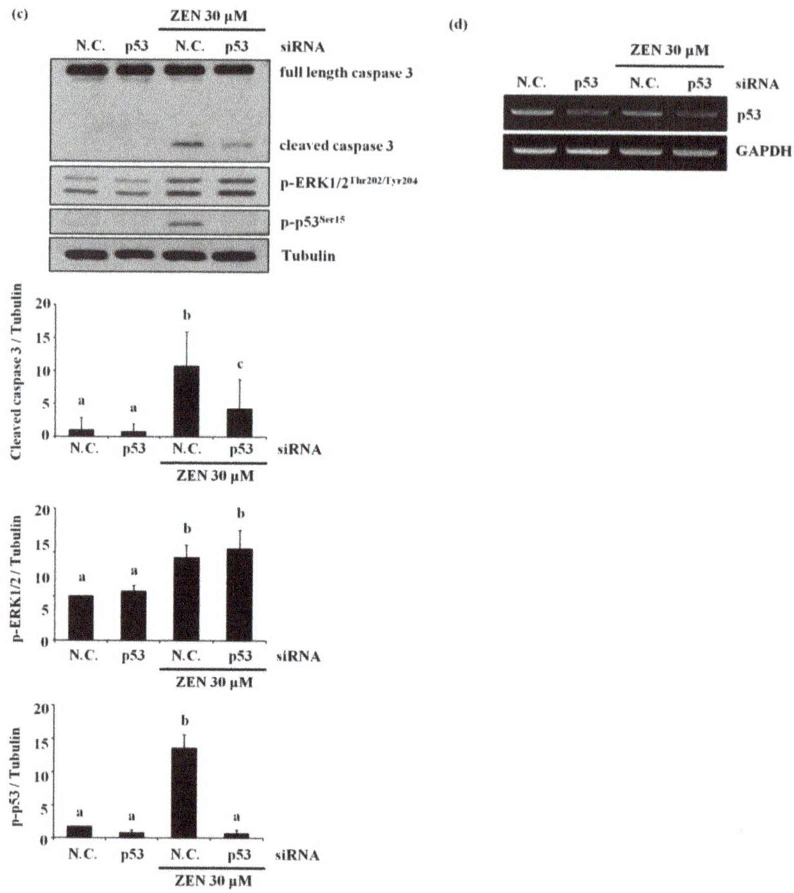

Figure 4. ZEN-induced EC apoptosis is mediated via a signaling axis of ERK1/2/p53/caspase 3. (**a**) After pretreatment with 1 μM of U0126 or 5 μM of pifithrin-α for 1 h, BAECs were incubated with 30 μM of ZEN for 24 h, and the protein expression of p-p53 and p-ERK1/2 relative to tubulin was quantified using western blot analyses. The plots depict the mean fold changes relative to the control (±SD) from at least four independent experimental trials. (**b**,**c**) The protein expression of cleaved caspase 3 and PARP relative to tubulin were quantified using western blot analyses in the BAECs transfected with siRNA of ERK1/2 and p53 with or without 30 μM of ZEN exposure for 24 h. (**d**) The mRNA expression of p53 relative to GAPDH in the BAECs were quantified using RT-PCR analyses. The plots depict the mean fold changes relative to the control (±SD) from at least four independent experimental trials. The different alphabetical letters refer to significant difference ($p < 0.05$) among groups, which were determined by one-way ANOVA followed by Tukey's multiple comparisons.

2.5. The Estrogen Receptor Is Not Involved in ZEN-Stimulated Apoptosis

Previously, it has been reported that ZEN exhibits some of its toxic effects through either genomic and/or nongenomic estrogen receptor-mediated signaling pathways [5,6]. Therefore, we examined whether the estrogen receptor was also involved in ZEN-induced apoptotic cell death. Treatment with ICI 182,780, a genomic estrogen receptor antagonist, did not alter cell viability and the levels of cleaved caspase 3 and PARP in BAECs treated with ZEN (Figure 5a,b). Similarly, G-15, a nongenomic estrogen receptor antagonist, also did not prevent ZEN-induced EC apoptosis (Figure 5c,d). Together, these results suggested that ZEN was not likely to induce apoptosis of BAECs through either the genomic or nongenomic estrogen receptor. As shown in Figures S1 and S2, we also confirmed that the concentrations and conditions of the estrogen receptor antagonists used in this experiment were effective enough to negate the estrogen-mediated response by ZEN; ICI 182,780 reversed the endothelial nitric oxide synthase (eNOS) mRNA expression induced by 17β-estradiol, and G-15 decreased the phosphorylation of eNOS at serine 1179 induced by G-1, a nongenomic estrogen receptor agonist, respectively.

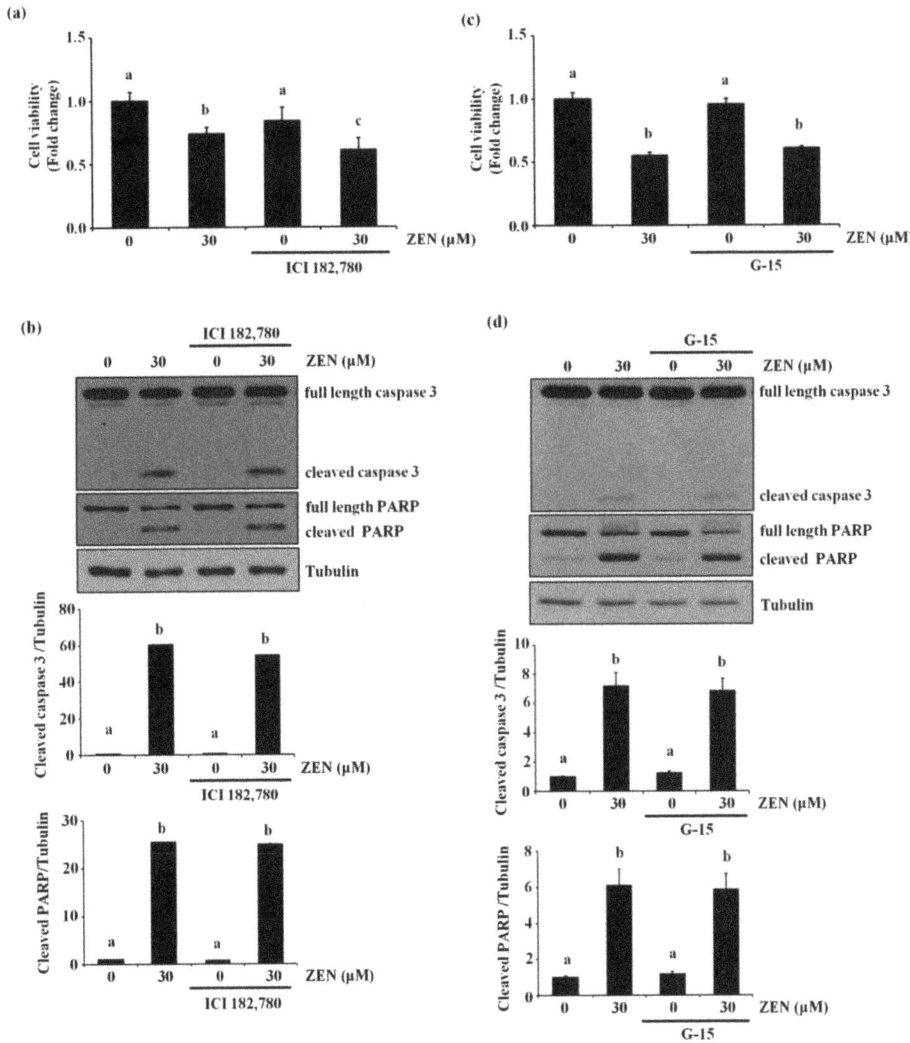

Figure 5. Estrogen receptors do not mediate ECs apoptosis by ZEN. After pretreatment with (**a**,**b**) 10 μM of ICI 182,780 or (**c**,**d**) 1 μM of G-15 for 1 h, BAECs were incubated with 30 μM of ZEN for 24 h. (**a**,**c**) Cell viability was measured using the MTT assay. (**b**,**d**) The protein expression of cleaved caspase 3 and PARP relative to tubulin was quantified using western blot analyses. The plots depict the mean fold changes relative to the control (±SD) from at least four independent experimental trials. The different alphabetical letters refer to significant differences ($p < 0.05$) among groups, which were determined by one-way ANOVA followed by Tukey's multiple comparisons.

2.6. ZEN Induces Reactive Oxygen Species (ROS), But ROS Are Not Involved in ZEN-Induced EC Apoptosis

Several studies showed that ZEN induces apoptosis by elevating intracellular ROS level [10,11]. To confirm whether ROS also plays a role in ZEN-induced apoptosis of ECs under our conditions, BAECs were treated with NAC, an antioxidant, followed by ZEN exposure. As shown in Figure 6a, ZEN significantly induced ROS production in BAECs, but ROS production induced by ZEN was prevented by NAC treatment. The NAC treatment did not reverse the effects of ZEN on cell viability and apoptosis (Figure 6b,c), indicating that ROS is unlikely to be responsible for the apoptotic death of BAECs induced by ZEN under our experimental conditions.

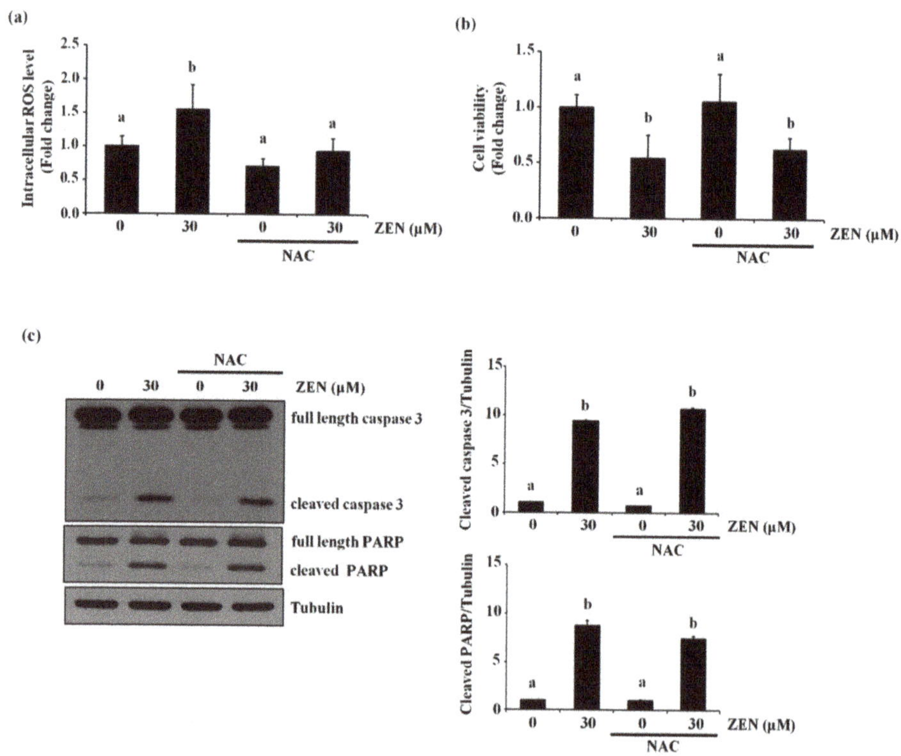

Figure 6. ZEN induces EC apoptosis in an ROS-independent manner. After pretreatment with 5 mM of NAC for 3 h, BAECs were incubated with 30 μM of ZEN for 24 h. (**a**) The intracellular ROS production from the BAECs was measured using DCF-DA. (**b**) Cell viability was measured by using the MTT assay. (**c**) The protein expression of cleaved caspase 3 and PARP relative to tubulin was quantified using western blot analyses. The plots depict the mean fold changes relative to the control (±SD) from at least four independent experimental trials. The different alphabetical letters refer to significant difference ($p < 0.05$) among groups, which were determined by one-way ANOVA followed by Tukey's multiple comparisons.

2.7. ZEN Mediates Apoptosis through a Cytosolic Ca^{2+}-Dependent Pathway

It has also been reported that ZEN increases cell death by increasing level of cytosolic Ca^{2+} [23,24]. Therefore, we examined whether ZEN induced EC apoptosis by increasing the level of cytosolic Ca^{2+} in BAECs. Pretreatment with the cytosolic Ca^{2+} chelator BAPTA-AM prior to ZEN exposure prevented ZEN-induced cell death of BAECs, as shown in Figure 7a. Furthermore, caspase 3 and PARP cleavage as well as ERK1/2 and p53 phosphorylation induced by ZEN were also reversed by BAPTA-AM treatment (Figure 7b). As shown in

Figure 7c, using confocal microscopy, we confirmed that the ZEN-induced cytosolic Ca^{2+} was reduced by BAPTA-AM. Interestingly, EGTA, an extracellular Ca^{2+} chelator, did not have the same effect as BAPTA-AM on the ZEN-treated BAECs (Figure 7d). From these results, we postulated that releasing cytosolic Ca^{2+} is likely the primary upstream event for triggering ERK1/2/p53/caspase 3-mediated apoptotic pathway by ZEN exposure.

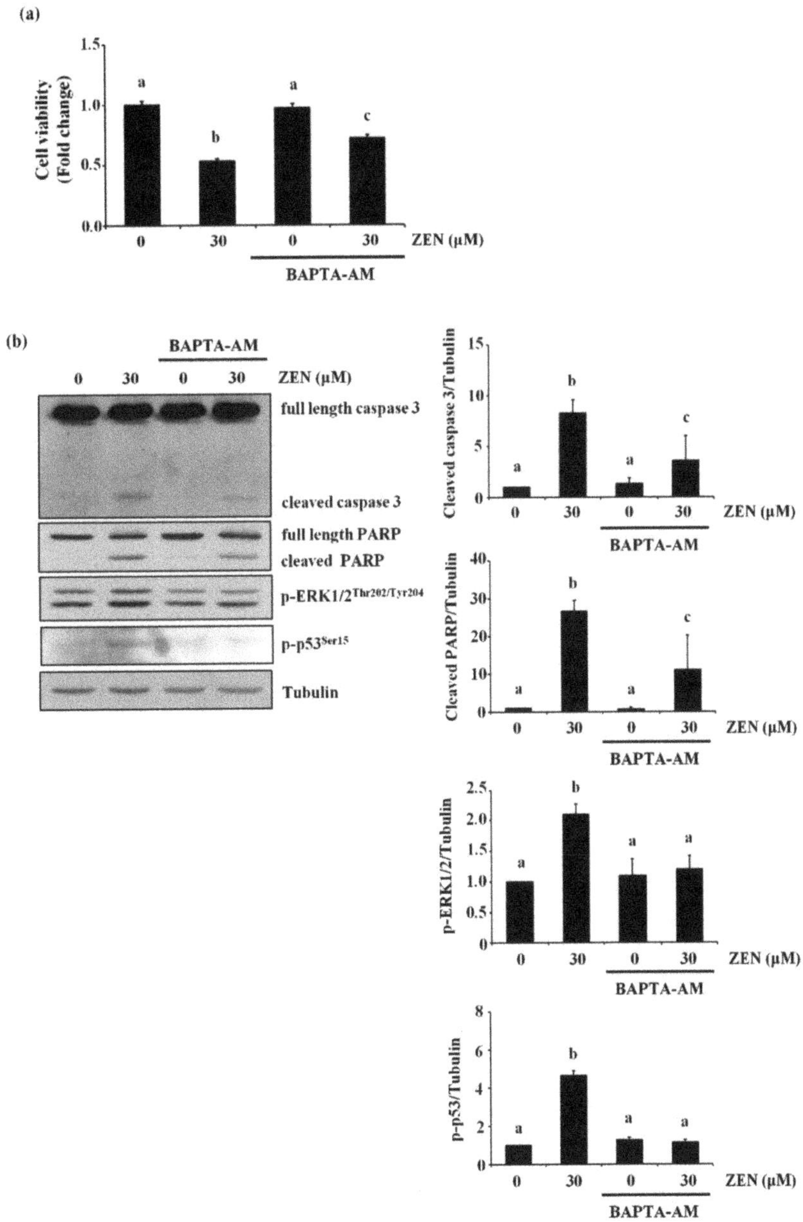

Figure 7. *Cont.*

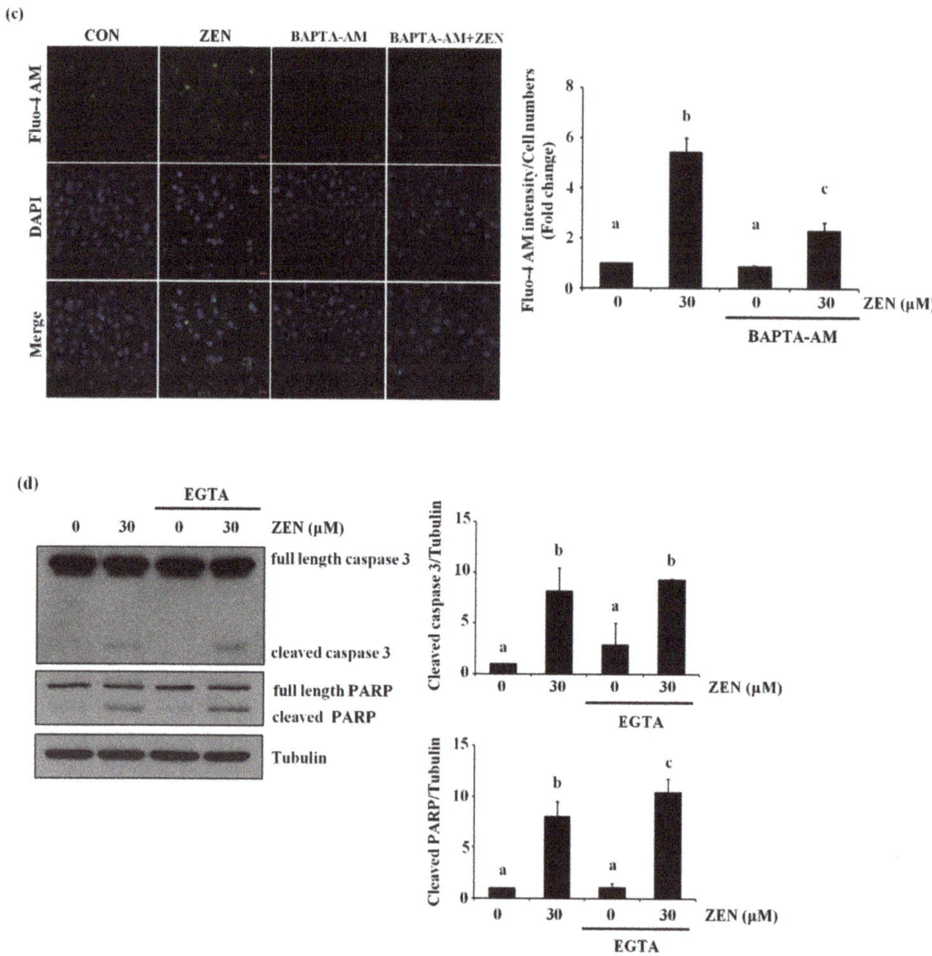

Figure 7. Cytosolic Ca^{2+} is involved in ZEN-induced EC apoptosis. After pretreatment with 2 μM of BAPTA-AM (**a**–**c**) or 0.1 mM of EGTA (**d**) for 1 h, BAECs were incubated with 30 μM of ZEN for 24 h. (**a**) Cell viability was measured using the MTT assay. (**b**,**d**) The protein expression of cleaved caspase 3 and PARP relative to tubulin was quantified using western blot analyses. The plots depict the mean fold changes relative to the control (±SD) from at least four independent experimental trials. (**c**) Visualization of cytosolic Ca^{2+} in the BAECs stained with the membrane-permeable Ca^{2+} indicator dye Fluo-4 AM using a confocal microscope (100×). The different alphabetical letters refer to significant difference ($p < 0.05$) among groups, which were determined by one-way ANOVA followed by Tukey's multiple comparisons.

2.8. The Endoplasmic Reticulum (ER) Is the Organelle Responsible for Cytosolic Ca^{2+} Release by ZEN

ZEN was reported to increase ER stress to induce apoptosis [12,25–27] and the ER is the major organelle responsible for storing and releasing Ca^{2+} [28]. Under our conditions, we examined whether ER stress was involved in the effects of ZEN on BAEC apoptosis. It has been known that when ER stress is induced, the phosphorylation of PERK is increased, which then activates the PERK/eIF2α-dependent pro-apoptotic transcriptional signal to induce apoptosis [29,30]. For this reason, we chose the level of phosphorylation of PERK as a marker of ER stress. As expected, ZEN increased the phosphorylation of PERK, which is decreased by pretreatment with 4-PBA, an ER stress inhibitor (Figure 8a). However, cleavage of caspase 3 and PARP induced by ZEN was not prevented with 4-PBA, indicating that ER stress is unlikely to be involved in the ZEN-induced apoptosis of BAECs. Lastly, we treated the BAECs with 2-APB, an ER Ca^{2+} channel inhibitor, and found that 2-APB significantly prevented the effects of ZEN on EC apoptosis (Figure 8b). Based on these findings, we concluded that ZEN induces BAEC apoptosis by promoting Ca^{2+} release from ER to elevate cytosolic Ca^{2+}, which subsequently activates ERK1/2/p53/caspase 3 apoptotic pathway.

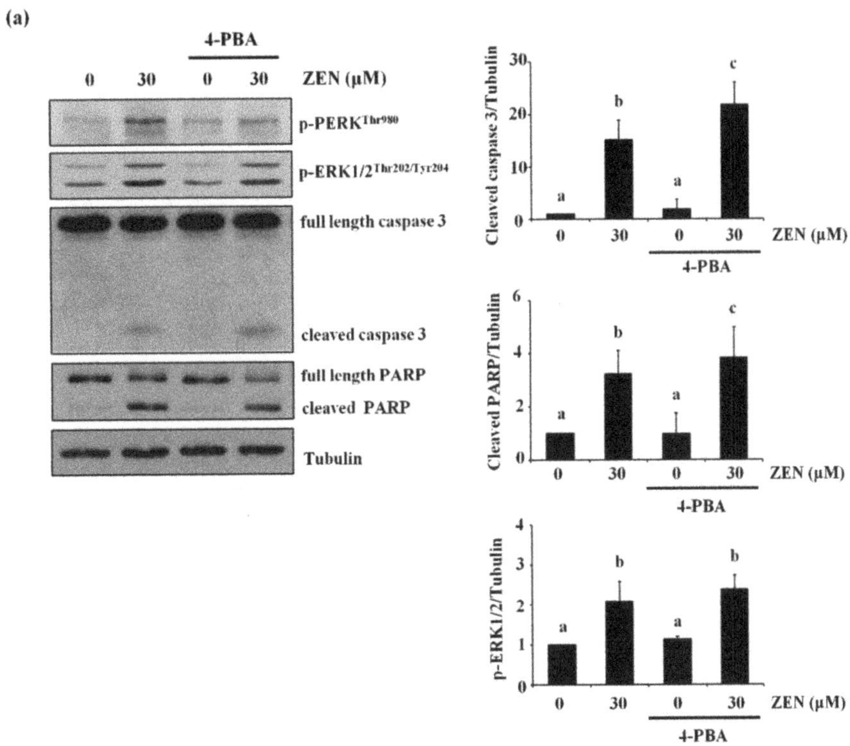

Figure 8. *Cont.*

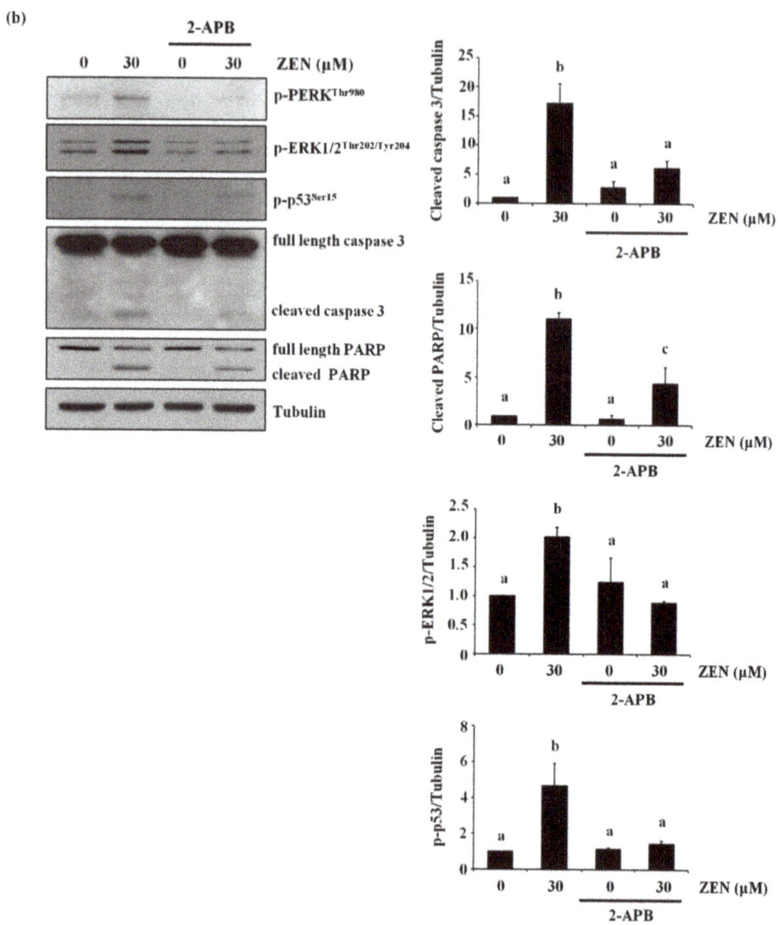

Figure 8. ER-mediated Ca^{2+} channel, but not ER stress, is involved in ZEN-induced apoptosis of BAECs. After pretreatment with (**a**) 2 mM of 4-PBA, an ER stress inhibitor, or (**b**) 20 μM of 2-APB, an ER-mediated Ca^{2+} channel inhibitor, for 1 h, BAECs were incubated with 30 μM of ZEN for 24 h. (**a**,**b**) The protein expression of cleaved caspase 3 and PARP relative to tubulin was quantified using western blot analyses. The plots depict the mean fold changes relative to the control (±SD) from at least four independent experimental trials. The different alphabetical letters refer to significant difference ($p < 0.05$) among groups, which were determined by one-way ANOVA followed by Tukey's multiple comparisons.

3. Discussion

Many studies have focused on the toxicological effects of ZEN on the cells/tissues related to reproductive and digestive function [10,31], but there have been only few studies addressing the effects on vascular function. It has been widely reported that ZEN alters the function and viability of various cells/tissues through interactions with the estrogen receptor [5,6], as well as by inducing ROS and ER stress [10,11,25,26]. Nevertheless, we found that ZEN induces apoptosis of BAECs through a mechanism that is independent of the estrogen receptor and ROS, but is closely associated with cytosolic Ca^{2+} levels and subsequent activation of an ERK1/2/p53/caspase 3 signaling axis. This study provides the first evidence of novel toxicological mechanism of ZEN-induced apoptosis using BAECs.

Our study demonstrated that ZEN disrupts vascular function by inducing apoptotic death of ECs, primarily through an ERK1/2/p53/caspase 3 signaling pathway (Figure 9).

ZEN was previously shown to induce apoptosis of RAW264.7 cells through a JNK and p38 MAPK pathway [14]. However, we found that ZEN induces apoptosis through an ERK1/2/p53/caspase 3 signaling pathway rather than through JNK and p38 MAPK. Similar to JNK and p38 MAPK, ERK1/2 belongs to the MAPK family, but ERK1/2 is better known as a signaling molecule that promotes cell proliferation. However, its role seems to be largely dependent on drugs and cell types, where it can also induce apoptosis, senescence and autophagy [26]. In this regard, various stimulants, such as DNA damaging agents (etoposide and UV) or anticancer compounds (resveratrol, taxol and oridonin), activate ERK1/2 to induce apoptosis in NIH 3T3, human papillary thyroid carcinoma, human melanoma and human breast adenocarcinoma cell lines [26]. Similar to these compounds, ZEN increases phosphorylation of ERK1/2 to further trigger other downstream pro-apoptotic molecules, including p53 and the caspase cascade of apoptosis, as shown in our study [32,33]. In some cases, p53 can be an activator of ERK1/2, initiating the p53/ERK1/2 signaling pathway [34,35]. However, given that inhibition of ERK1/2 decreased ZEN-stimulated phosphorylation of p53 but not *vice versa*, it is likely that ERK1/2 acts as upstream of p53 to initiate apoptosis of ECs, suggesting that ZEN-induced apoptosis occurs through sequential activation of ERK1/2, p53 and caspase 3.

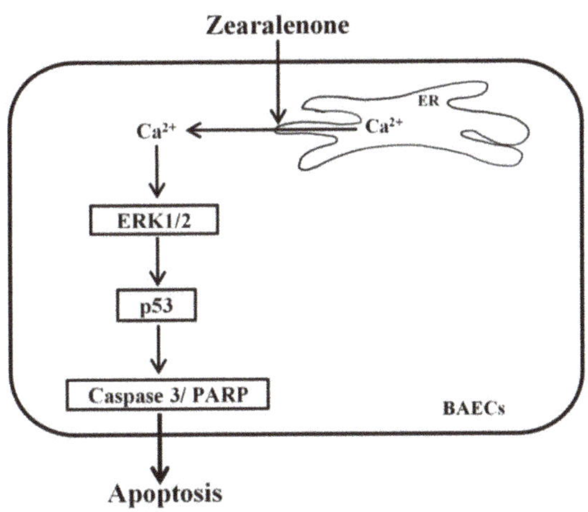

Figure 9. Schematic illustration of the molecular mechanism by which ZEN induces apoptosis in BAECs. ZEN increases the levels of cytosolic Ca^{2+} that are released from ER through ER Ca^{2+} channel activation. The increased cytosolic Ca^{2+} levels stimulate the phosphorylation of ERK1/2 and p53, subsequently enhancing the cleavage of caspase 3 and PARP and resulting in apoptosis of BAECs.

Several factors are involved in the activation of the ERK1/2/p53/caspase pathway. For example, cytosolic Ca^{2+} release activates molecules such as Ras, $G_{\alpha i2}$, calpain, and calmodulin kinase 1, which then regulate ERK1/2 activity [36–38]. In this regard, a previous study reported that calcimycin activates Ras through Ras-guanine nucleotide releasing factor 2, consequently triggering the cytosolic Ca^{2+}/ERK1/2/p53 signaling pathway to induce apoptosis of the rabbit lens epithelial cells [39]. Our study also validates the findings of this previous study, where chelation of cytosolic Ca^{2+} using BAPTA-AM, but not extracellular Ca^{2+}, effectively prevents apoptosis as well as the expression of pro-apoptotic molecules induced by ZEN. These findings demonstrate an important role for the influx of cytosolic Ca^{2+} in inducing the ERK/1/2/p53/caspase 3-mediated apoptotic signaling pathway in ECs exposed to ZEN. Furthermore, our results also indicate that the ZEN-induced increase in cytosolic Ca^{2+} results from the activation of Ca^{2+} channels located in the ER. These findings are fairly consistent with those of the recent study showing that compound K, a major metabolite of ginsenosides, induces apoptosis by activating ER Ca^{2+}

channels known as ryanodine receptors to release ER Ca^{2+} in human lung cancer cells, the A549, and SK-MES-1 cell lines [40]. Although ZEN has also been reported to increase cytosolic Ca^{2+} level by inducing ER stress in TM4 cells [41], ovarian cells from pre-pubertal bitches [42] and lymphocytes of chickens [24], our data do not support the involvement of ER stress in ZEN-induced apoptosis. At present, the reason for these inconsistencies has not been determined, but they may be attributable to the different cell types used in our study; BAECs versus TM4 and ovarian cells.

Many studies have reported that ZEN generates ROS, which in turn induces DNA damage and cell death. For example, induction of ROS and malondialdehyde production, as well as loss of mitochondrial membrane potential by ZEN leads to apoptotic death of SIEC02 and SHSY-5Y cells [10,11]. We also found increased ROS production with ZEN treatment, but complete inhibition of ROS production by NAC had no effect on the apoptosis induced by ZEN, as shown in Figure 6. Consistent with our findings, in a study examining DNA damage as an indicator of apoptosis in human embryonic kidney cell line HEK293 cells, ZEN-induced cell death is found to be mediated through a ROS-independent pathway [8], and blocking of ROS using the antioxidant hydroxytyrosol did not prevent the ZEN-induced DNA damage. The authors concluded that lysosomal injury is a potential cause of DNA damage and apoptosis induced by ZEN. Other ROS inducers, such as artesunate and dihydroartemisinin, are also known to mediate apoptotic cell death through a ROS-independent pathway; the former induces Bax-mediated apoptosis in HepG2 cells [43], whereas the latter induces apoptosis through activation of the p38 MAPK/caspase 9/caspase 3 signaling pathway in leukemia HL-60 cells [44].

ZEN has long been known to be an estrogenic mycotoxin that competitively binds to the estrogen receptors to alter the synthesis and secretion of steroidal reproductive hormones [4] as well as sperm quality of rats by increasing apoptosis and necrosis [5]. Therefore, we examined the potential involvement of genomic and nongenomic estrogen receptors in the ZEN-induced apoptotic death of BAECs. Under our experimental conditions, however, ZEN-mediated apoptosis of BAECs was not prevented by adding the inhibitors of either genomic or nongenomic estrogen receptor, as shown in Figure 5. These findings indicate that estrogenic activity mediated by ZEN is unlikely to induce apoptosis, at least for ECs.

4. Conclusions

In conclusion, our study is the first to demonstrate that ZEN induces EC apoptosis by activating a signaling axis of cytosolic Ca^{2+}/ERK1/2/p53, as illustrated in Figure 9. This molecular mechanism represents a potential risk of ZEN exposure in relation to vascular health in animals and humans.

5. Materials and Methods

5.1. Materials

ZEN, ICI 182,780 (genomic estrogen receptor antagonist), N-acetyl-L-cysteine (NAC; antioxidant), sodium phenylbutyrate (4-PBA; endoplasmic reticulum (ER) stress inhibitor), BAPTA-AM (cytosolic calcium chelator), U0126 (ERK inhibitor), pifithrin-α (p53 inhibitor), SP600125 (JNK inhibitor), SB203580 (p38 MAPK inhibitor), 2′,7′-dichlorofluorescin-diacetate (DCF-DA) and ethylene glycol-*bis*(2-aminoethylether)-*N*,*N*,*N*′,*N*′-tetraacetic acid (EGTA; extracellular Ca^{2+} chelator) were purchased from Merck (Darmstadt, Germany). Z-DVED-FMK (caspase 3 inhibitor) and G-15 (nongenomic estrogen receptor antagonist) were purchased Tocris bioscience (Bristol, UK). Antibodies against ERK1/2 and all corresponding secondary antibodies were obtained from Santa Cruz Biotechnology (Dallas, TX, USA). Antibody against α-tubulin was purchased from GW Vitek (Seoul, Korea), Antibodies against caspase 3, poly ADP-ribose polymerase (PARP), phosphorylated p44/42 MAPK (ERK1/2) (p-ERK1/$2^{Thr202/Tyr204}$), p-$JNK^{Thr183/Tyr15}$, JNK, p-MAP kinase-activated protein kinase (MAPKAPK)-Thr334 (p-$MAPKAPK^{Thr334}$), MAPKAPK, p-$p53^{Ser15}$ and p-$PERK^{Thr980}$ were purchased from Cell Signaling Technology (Danvers, MA, USA). Lipofectamine 2000, minimal essential medium (MEM), Dulbecco's phosphate-buffered saline (DPBS), newborn calf serum (NCS), penicillin-streptomycin antibiotics, L-glutamine, trypsin-EDTA solution and plasticware for cell culture were obtained from Thermo Fisher Scientific (Waltham, MA, USA). All other chemicals were of the purest analytical grade.

5.2. Cell Culture and Drug Treatments

BAECs were isolated and cultured as described previously [45] and maintained in MEM supplemented with 5% NCS at 37 °C under 5% CO_2. Cells between passages 5 and 9 were used in 1% NCS for all experiments. The cells were incubated with ZEN at various concentrations for 24 h or with 30 μM ZEN for 4, 8, 16 or 24 h. In separate experiments, various chemicals were added to the cell cultures 1 or 3 h prior to ZEN treatment.

5.3. Cell Viability Assay

The cell viability assay was carried out as described [46] using 3-(4,5-dimethylthiazol-2-yl)-2,5-diphenyltetrazolium bromide (MTT, Merck) with minor modifications. BAECs were seeded at a density of 1.0×10^4 per well in a 96-well culture plate (four replicates for each treatment), and incubated with ZEN at various concentrations (0, 10, 30 or 60 μM) for 24 h or with 30 μM ZEN for various time points (4, 8, 16 or 24 h). After the ZEN treatments, the cells were incubated with 5 mg/mL MTT and further incubated for 1 h at 37 °C. The cells were then treated with dimethylsulfoxide (DMSO) for 10 min, and the absorbance was read at 570 nm using a 96-well microtiter plate reader (BioTek Instruments, Winooski, VT, USA).

5.4. Annexin V-Fluorescein Isothiocynate (FITC)/Propidium Iodide (PI) Double Staining

Apoptosis of BAECs was measured using annexin V-FITC/PI apoptosis detection kit (BD Biosciences, Franklin Lakes, NJ, USA). Briefly, cells were trypsinized and harvested after centrifugation at $2000\times g$ for 3 min, then re-suspended in 500 μL of binding buffer. Annexin-FITC (5 μL) and PI (5 μL) were to the harvested cells in solution and incubated for 15 min at 37 °C in the dark. Scattering signals were detected by fluorescence-activated single cell sorting (FACS) analysis using a flow cytometer (ACEA Biosciences, San Diego, CA, USA). In this study, the percentage of apoptotic cell population was assessed from the annexin V+/PI− population that represents early apoptotic cells.

5.5. Detection of Intracellular ROS Levels

Intracellular ROS was measured using the oxidation-sensitive fluorescent probe DCF-DA (Merck), in an assay based on the ROS-dependent oxidation of DCF-DA to 2′,7′-dichlorofluorescein (DCF), as described previously [47] with minor modifications. Briefly, BAECs grown in 96-well culture plates were incubated with 20 μM DCF-DA for 30 min. The DCF-DA solution was removed and the cells were washed with DPBS. The cells were then incubated with NAC for 3 h before ZEN treatment and incubated for an additional 24 h at 37 °C under 5% CO_2 after the addition of ZEN. The deposited intracellular DCF-DA was measured using a 96-well microtiter plate reader (ex 485 nm/em 530 nm; BioTek Instrument).

5.6. Small Interfering RNA (siRNA) Transfection

BAECs were transfected with 100 nM of small interfering RNA (siRNA) using Dharmafect 4 (Dharmacon, Lafayette, CO, USA) according to the manufacturer's instructions. siRNA oligonucleotides for ERK1, ERK2, p53 and negative control siRNA were purchased from GenePharma (Shanghai, China) as follows: *ERK1*, CCU UUG AGC AUC AGA CCU ACU, CCA AGG AAC GAC UGA AGG AGC; *ERK2*, AGA AAA UCA GCC CUU UUG AGC, AAA UCA UGU UGA AUU CCA AGG; *p53*, ACU ACA AUU UCA UGU GUA ACA, GGU UUA AAC GCU AUG AGA UGU, CAU ACA CUG GGU UGG AAA ACU; and the negative control (N.C.), UUC UCC GAA CGU GUC ACG UTT. At least two sequences of siRNA oligonucleotides were designed and combined to increase the efficiency of knockdown of the gene of interest. The BAECs with or without transfections were grown to confluence and maintained for an additional 24 h in MEM with 1% NCS containing 30 μM ZEN.

5.7. Western Blot Analyses

For the western blot analyses, BAECs treated with ZEN in the absence or presence of various additional chemicals were washed with ice-cold Dulbecco's phosphate-buffered saline (DPBS) and lysed with lysis buffer (20 mM Tris-HCl at pH 7.5, 150 mM NaCl, 1% Triton X-100, 1 mM EDTA, 1 mM EGTA) containing Protease Inhibitor Cocktail™ (Merck), 1 mM β-glycerophosphate, 1 mM phenylmethanesulfonyl fluoride, 1 mM NaF and 1 mM Na_3VO_4. The protein concentrations were determined using a BCA protein assay (Merck). Equal quantities of protein (20 μg) were separated on sodium dodecyl sulfate polyacrylamide gel under reducing conditions and then electrophoretically transferred onto nitrocellulose membranes. The blots were then probed with appropriate antibodies, each at a 1:1000 dilution, followed by the corresponding secondary antibodies, and finally developed using enhanced chemiluminescence reagents (ECL, GE Healthcare, Chicago, IL, USA). Proteins on the nitrocellulose membranes were quantified using Image J software (National Institutes of Health, Bethesda, MD, USA). The tubulin was used as a loading control to normalize the quantified values of target proteins of interest.

5.8. Reverse Transcription-Polymerase Chain Reaction (PCR)

Total RNA was extracted from BAECs using TRIzol™ Reagent (Thermo Fisher Scientific) as described previously [48]. Briefly, the cells were homogenized in 1 mL of TRIzol™ reagent. The total RNA was then converted to cDNA using SuperScript™ III reverse transcriptase (Thermo Fisher Scientific). PCR amplification of a cDNA encoding each target gene was conducted using the following primers: *p53*-F, 5′-CTA CCA ACA CCA GCT CCT-3′; *p53*-R, 5′-CGG CTC ACA GTA AAA ACC TT-3′; *GAPDH*-F, 5′-TCA CCA GGG CTG CTT TTA AT-3′; *GAPDH*-R, 5′-GGT CAT AAG TCC CTC CAC GA-3′. The amplified products were separated using a 1% agarose gel in TAE buffer (40 mM Tris-acetate, pH 8.0, 1 mM EDTA). The band of the target gene was visualized under UV light using RedSafe™ Nucleic acid Staining Solution (iNtRON biotechnology, Gyeonggi-do, Korea).

5.9. Measurement of Cytosolic Ca^{2+} Levels

Cytosolic Ca^{2+} levels were measured using the membrane-permeable Ca^{2+} indicator dye Fluo-4 AM (Thermo Fisher Scientific), as described previously [49]. BAECs were pretreated with BAPTA-AM for 1 h before ZEN treatment in the presence of 1 μM Fluo-4 AM. Fluorescent images were obtained using a confocal microscope (LSM5 Pascall, Carl ZEISS, Oberkochen, Germany). The cytosolic Ca^{2+} level was quantified by measuring the fluorescence intensity (Green) using Image J software, and the number of cells from each image was estimated by counting the nucleus stained with DAPI (Blue). The fluorescence intensity of cytosolic Ca^{2+} level was then normalized by dividing the counted cell numbers.

5.10. Statistical Analyses

All statistical analyses were performed using GraphPad Prism software ver. 5 (GraphPad Software, San Diego, CA). Data are expressed as the mean ± standard deviation (SD) and statistical significance ($p < 0.05$) was determined using one-way ANOVA followed by Tukey's multiple comparison test. The significant differences are denoted with different alphabetical letters.

Supplementary Materials: The following are available online at https://www.mdpi.com/2072-6651/13/3/187/s1. Figure S1. The inhibitory effect of ICI182,780 on eNOS mRNA expression in BAECs. After pretreatment with 10 μM of ICI182,780 for 1 h, BAECs were incubated with 1 nM 17β-estradiol for 24 h. The eNOS mRNA expression was quanti-fied using qRT-PCR. The plots depict the mean fold changes relative to control (±SD) from at least three independent experimental trials. The different alpha-betical letters refer to significant differences ($p < 0.05$) among groups, which were determined by one-way ANOVA followed by Tukey's multiple comparison, Figure S2. The inhibitory effect of G-15 on the expression of total eNOS and p-eNOSSer1179 in BAECs. After pretreatment with 1 μM of G-15 for 1 h, BAECs were incubated with 100 nM G-1 for 24 h. The protein expression of p-eNOSSer1179 relative to eNOS was quantified using Western blot analyses. The plots depict the mean fold changes relative to control (±SD) from at least three independent experimental trials. The different alphabetical letters refer to significant differ-ences ($p < 0.05$) among groups, which were determined by one-way ANOVA followed by Tukey's multiple comparison.

Author Contributions: Conceptualization, H.-J.L. and I.J.; Methodology, H.-J.L.; Software, H.-J.L.; Validation, H.-J.L.; Formal analysis, H.-J.L.; Investigation, H.-J.L.; Resources, H.-J.L. and I.J.; Data curation, H.-J.L.; Writing—original draft, H.-J.L.; Writing—review & editing, I.J. and S.-Y.O.; Visualization, H.-J.L.; Supervision, I.J.; Project administration, I.J.; Funding acquisition, S.-Y.O. and I.J. All authors have read and agreed to the published version of the manuscript.

Funding: This work was supported by the Basic Science Research Program [NRF-2017M3A9B3063636 and NRF-2018R1A2B2002062] and Small Grant for Exploratory Research [NRF-2018R1D1A1A02085696] programs through the National Research Foundation of Korea (NRF) funded by the Ministry of Science, ICT, and Future Planning.

Institutional Review Board Statement: Not applicable.

Informed Consent Statement: Not applicable.

Data Availability Statement: Not applicable.

Conflicts of Interest: All authors declare they have no conflict of interest.

References

1. Hossain, Z.; Mari, N.; Goto, T. The relationship between ergosterol and mycotoxin contamination in maize from various countries. *Mycotoxin Res.* **2015**, *31*, 91–99. [CrossRef] [PubMed]
2. Belhassen, H.; Jimenez-Diaz, I.; Ghali, R.; Ghorbel, H.; Molina-Molina, J.M.; Olea, N.; Hedili, A. Validation of a UHPLC-MS/MS method for quantification of zearalenone, alpha-zearalenol, beta-zearalenol, alpha-zearalanol, be-ta-zearalanol and zearalanone in human urine. *J. Chromatogr. B* **2014**, *962*, 68–74. [CrossRef] [PubMed]
3. Rogowska, A.; Pomastowski, P.; Sagandykova, G.; Buszewski, B. Zearalenone and its metabolites: Effect on human health, metabolism and neutralisation methods. *Toxicon* **2019**, *162*, 46–56. [CrossRef] [PubMed]
4. Zheng, W.; Feng, N.; Wang, Y.; Noll, L.; Xu, S.; Liu, X.; Lu, N.; Zou, H.; Gu, J.; Yuan, Y.; et al. Effects of zearalenone and its derivatives on the synthesis and secretion of mammalian sex steroid hormones: A review. *Food Chem. Toxicol.* **2019**, *126*, 262–276. [CrossRef]
5. Adibnia, E.; Razi, M.; Malekinejad, H. Zearalenone and 17 beta-estradiol induced damages in male rats reproduction potential; evidence for ERalpha and ERbeta receptors expression and steroidogenesis. *Toxicon* **2016**, *120*, 133–146. [CrossRef] [PubMed]
6. He, J.; Wei, C.; Li, Y.; Liu, Y.; Wang, Y.; Pan, J.; Liu, J.; Wu, Y.; Cui, S. Zearalenone and alpha-zearalenol inhibit the synthesis and secretion of pig follicle stimulating hormone via the non-classical estrogen membrane receptor GPR30. *Mol. Cell. Endocrinol.* **2018**, *461*, 43–54. [CrossRef]
7. Zatecka, E.; Ded, L.; Elzeinova, F.; Kubátová, A.; Dorosh, A.; Margaryan, H.; Dostalova, P.; Korenkova, V.; Hošková, K.; Pěknicová, J. Effect of zearalenone on reproductive parameters and expression of selected testicular genes in mice. *Reprod. Toxicol.* **2014**, *45*, 20–30. [CrossRef]
8. Gao, F.; Jiang, L.-P.; Chen, M.; Geng, C.-Y.; Yang, G.; Ji, F.; Zhong, L.-F.; Liu, X.-F. Genotoxic effects induced by zearalenone in a human embryonic kidney cell line. *Mutat. Res. Toxicol. Environ. Mutagen.* **2013**, *755*, 6–10. [CrossRef]
9. Lu, J.; Yu, J.Y.; Lim, S.S.; Son, Y.O.; Kim, D.H.; Lee, S.A.; Shi, X.; Lee, J.C. Cellular mechanisms of the cytotoxic effects of the zearalenone metabolites alpha-zearalenol and beta-zearalenol on RAW264 7 macrophages. *Toxicol. In Vitro* **2013**, *27*, 1007–1017. [CrossRef]
10. Zheng, W.-L.; Wang, B.-J.; Wang, L.; Shan, Y.-P.; Zou, H.; Song, R.-L.; Wang, T.; Gu, J.-H.; Yuan, Y.; Liu, X.-Z.; et al. ROS-Mediated Cell Cycle Arrest and Apoptosis Induced by Zearalenone in Mouse Sertoli Cells via ER Stress and the ATP/AMPK Pathway. *Toxins* **2018**, *10*, 24. [CrossRef] [PubMed]

11. Venkataramana, M.; Nayaka, S.C.; Anand, T.; Rajesh, R.; Aiyaz, M.; Divakara, S.; Murali, H.; Prakash, H.; Rao, P.L. Zearalenone induced toxicity in SHSY-5Y cells: The role of oxidative stress evidenced by N-acetyl cysteine. *Food Chem. Toxicol.* **2014**, *65*, 335–342. [CrossRef]
12. Lin, P.; Chen, F.; Sun, J.; Zhou, J.; Wang, X.; Wang, N.; Li, X.; Zhang, Z.; Wang, A.; Jin, Y. Mycotoxin zearalenone induces apoptosis in mouse Leydig cells via an endoplasmic reticulum stress-dependent signalling pathway. *Reprod. Toxicol.* **2015**, *52*, 71–77. [CrossRef]
13. Wang, Y.; Zheng, W.; Bian, X.; Yuan, Y.; Gu, J.; Liu, X.; Liu, Z.; Bian, J. Zearalenone induces apoptosis and cytoprotective autophagy in primary Leydig cells. *Toxicol. Lett.* **2014**, *226*, 182–191. [CrossRef]
14. Yu, J.-Y.; Zheng, Z.-H.; Son, Y.-O.; Shi, X.; Jang, Y.-O.; Lee, J.-C. Mycotoxin zearalenone induces AIF- and ROS-mediated cell death through p53- and MAPK-dependent signaling pathways in RAW264.7 macrophages. *Toxicol. Vitr.* **2011**, *25*, 1654–1663. [CrossRef]
15. Anderson, C.; Majeste, A.; Hanus, J.; Wang, S. E-Cigarette Aerosol Exposure Induces Reactive Oxygen Species, DNA Damage, and Cell Death in Vascular Endothelial Cells. *Toxicol. Sci.* **2016**, *154*, 332–340. [CrossRef] [PubMed]
16. Ouyang, J.; Li, R.; Shi, H.; Zhong, J. Curcumin Protects Human Umbilical Vein Endothelial Cells against H_2O_2-Induced Cell Injury. *Pain Res. Manag.* **2019**, *2019*, 3173149. [CrossRef]
17. Ahmad, S.; Alam Khan, S.; Kindelin, A.; Mohseni, T.; Bhatia, K.; Hoda, N.; Ducruet, A.F. Acetyl-11-keto-β-boswellic acid (AKBA) Attenuates Oxidative Stress, Inflammation, Complement Activation and Cell Death in Brain Endothelial Cells Following OGD/Reperfusion. *Neuro Mol. Med.* **2019**, *21*, 505–516. [CrossRef]
18. Gajecki, M. Zearalenone—Undesirable substances in feed. *Pol. J. Veter-Sci.* **2002**, *5*, 117–122.
19. Zheng, W.; Wang, B.; Li, X.; Wang, T.; Zou, H.; Gu, J.; Yuan, Y.; Liu, X.; Bai, J.; Bian, J.; et al. Zearalenone Promotes Cell Proliferation or Causes Cell Death? *Toxins* **2018**, *10*, 184. [CrossRef]
20. Wang, H.; Zhao, X.; Ni, C.; Dai, Y.; Guo, Y. Zearalenone regulates endometrial stromal cell apoptosis and migration via the promotion of mitochondrial fission by activation of the JNK/Drp1 pathway. *Mol. Med. Rep.* **2018**, *17*, 7797–7806. [CrossRef] [PubMed]
21. Pistol, G.C.; Braicu, C.; Motiu, M.; Gras, M.A.; Marin, D.E.; Stancu, M.; Calin, L.; Israel-Roming, F.; Berindan-Neagoe, I.; Taranu, I. Zearalenone mycotoxin affects immune mediators, MAPK signalling molecules, nuclear receptors and ge-nome-wide gene expression in pig spleen. *PLoS ONE* **2015**, *10*, e0127503. [CrossRef] [PubMed]
22. Ayed-Boussema, I.; Bouaziz, C.; Rjiba, K.; Valenti, K.; Laporte, F.; Bacha, H.; Hassen, W. The mycotoxin Zearalenone induces apoptosis in human hepatocytes (HepG2) via p53-dependent mitochondrial signaling pathway. *Toxicol. Vitr.* **2008**, *22*, 1671–1680. [CrossRef]
23. Jilani, K.; Lang, F. Ca2+-dependent suicidal erythrocyte death following zearalenone exposure. *Arch. Toxicol.* **2013**, *87*, 1821–1828. [CrossRef]
24. Wang, Y.C.; Deng, J.L.; Xu, S.W.; Peng, X.; Zuo, Z.C.; Cui, H.M.; Ren, Z.H.; Wang, Y. Effects of zearalenone on calcium homeostasis of splenic lymphocytes of chickens in vitro. *Poult. Sci.* **2012**, *91*, 1956–1963. [CrossRef]
25. Wang, M.; Kern, A.M.; Hülskötter, M.; Greninger, P.; Singh, A.; Pan, Y.; Chowdhury, D.; Krause, M.; Baumann, M.; Benes, C.H.; et al. EGFR-Mediated Chromatin Condensation Protects KRAS-Mutant Cancer Cells against Ionizing Radiation. *Cancer Res.* **2014**, *74*, 2825–2834. [CrossRef] [PubMed]
26. Cagnol, S.; Chambard, J.C. ERK and cell death: Mechanisms of ERK-induced cell death—Apoptosis, autophagy and se-nescence. *FEBS J.* **2010**, *277*, 2–21. [CrossRef]
27. Chen, F.; Li, Q.; Zhang, Z.; Lin, P.; Lei, L.; Wang, A.; Jin, Y. Endoplasmic Reticulum Stress Cooperates in Zeara-lenone-Induced Cell Death of RAW 264.7 Macrophages. *Int. J. Mol. Sci.* **2015**, *16*, 19780–19795. [CrossRef] [PubMed]
28. Celli, A.; Crumrine, D.; Meyer, J.M.; Mauro, T.M. Endoplasmic Reticulum Calcium Regulates Epidermal Barrier Response and Desmosomal Structure. *J. Investig. Dermatol.* **2016**, *136*, 1840–1847. [CrossRef]
29. Sano, R.; Reed, J.C. ER stress-induced cell death mechanisms. *Biochim. Biophys. Acta (BBA) Bioenerg.* **2013**, *1833*, 3460–3470. [CrossRef] [PubMed]
30. Romine, I.C.; Wiseman, R.L. PERK Signaling Regulates Extracellular Proteostasis of an Amyloidogenic Protein During Endoplasmic Reticulum Stress. *Sci. Rep.* **2019**, *9*, 1–11. [CrossRef]
31. Zhu, L.; Yuan, H.; Guo, C.; Lu, Y.; Deng, S.; Yang, Y.; Wei, Q.; Wen, L.; He, Z. Zearalenone induces apoptosis and necrosis in porcine granulosa cells via a caspase-3- and caspase-9-dependent mitochondrial signaling pathway. *J. Cell. Physiol.* **2012**, *227*, 1814–1820. [CrossRef] [PubMed]
32. Wang, Y.; Xu, H.; Lu, Z.; Yu, X.; Lv, C.; Tian, Y.; Sui, D. Pseudo-Ginsenoside Rh2 induces A549 cells apoptosis via the Ras/Raf/ERK/p53 pathway. *Exp. Ther. Med.* **2018**, *15*, 4916–4924. [CrossRef] [PubMed]
33. Yang, S.-H.; Wang, S.-M.; Syu, J.-P.; Chen, Y.; Wang, S.-D.; Peng, Y.-S.; Kuo, M.-F.; Kung, H.-N. Andrographolide Induces Apoptosis of C6 Glioma Cells via the ERK-p53-Caspase 7-PARP Pathway. *BioMed Res. Int.* **2014**, *2014*, 1–15. [CrossRef] [PubMed]
34. Lee, S.Y.; Shin, S.J.; Kim, H.S. ERK1/2 activation mediated by the nutlin3induced mitochondrial translocation of p53. *Int. J. Oncol.* **2013**, *42*, 1027–1035. [CrossRef]
35. Lee, S.Y.; Choi, H.C.; Choe, Y.J.; Shin, S.J.; Lee, S.H.; Kim, H.S. Nutlin-3 induces BCL2A1 expression by activating ELK1 through the mitochondrial p53-ROS-ERK1/2 pathway. *Int. J. Oncol.* **2014**, *45*, 675–682. [CrossRef]
36. Holstein, D.M.; Berg, K.A.; Leeb-Lundberg, L.M.; Olson, M.S.; Saunders, C. Calcium-sensing receptor-mediated ERK1/2 activation requires Galphai2 coupling and dynamin-independent receptor internalization. *J. Biol. Chem.* **2004**, *279*, 10060–10069. [CrossRef]

37. Veeranna; Kaji, T.; Boland, B.; Odrljin, T.; Mohan, P.; Basavarajappa, B.S.; Peterhoff, C.; Cataldo, A.; Rudnicki, A.; Amin, N.; et al. Calpain mediates calcium-induced activation of the erk1,2 MAPK pathway and cytoskeletal phosphorylation in neurons: Relevance to Alzheimer's disease. *Am. J. Pathol.* **2004**, *165*, 795–805.
38. Schmitt, J.M.; Wayman, G.A.; Nozaki, N.; Soderling, T.R. Calcium Activation of ERK Mediated by Calmodulin Kinase I. *J. Biol. Chem.* **2004**, *279*, 24064–24072. [CrossRef]
39. Li, D.W.-C.; Liu, J.-P.; Mao, Y.-W.; Xiang, H.; Wang, J.; Ma, W.-Y.; Dong, Z.; Pike, H.M.; Brown, R.E.; Reed, J.C. Calcium-activated RAF/MEK/ERK Signaling Pathway Mediates p53-dependent Apoptosis and Is Abrogated by αB-Crystallin through Inhibition of RAS Activation. *Mol. Biol. Cell* **2005**, *16*, 4437–4453. [CrossRef]
40. Shin, D.-H.; Leem, D.-G.; Shin, J.-S.; Kim, J.-I.; Kim, K.-T.; Choi, S.Y.; Lee, M.-H.; Choi, J.-H.; Lee, K.-T. Compound K induced apoptosis via endoplasmic reticulum Ca2+ release through ryanodine receptor in human lung cancer cells. *J. Ginseng Res.* **2018**, *42*, 165–174. [CrossRef] [PubMed]
41. Feng, N.; Wang, B.; Cai, P.; Zheng, W.; Zou, H.; Gu, J.; Yuan, Y.; Liu, X.; Liu, Z.; Bian, J. ZEA-induced autophagy in TM4 cells was mediated by the release of Ca(2+) activates CaMKKbeta-AMPK signaling pathway in the endoplasmic reticulum. *Toxicol. Lett.* **2020**, *323*, 1–9. [CrossRef] [PubMed]
42. Gajęcka, M. The effect of low-dose experimental zearalenone intoxication on the immunoexpression of estrogen receptors in the ovaries of pre-pubertal bitches. *Pol. J. Veter-Sci.* **2012**, *15*, 685–691. [CrossRef]
43. Qin, G.; Wu, L.; Liu, H.; Pang, Y.; Zhao, C.; Wu, S.; Wang, X.; Chen, T. Artesunate induces apoptosis via a ROS-independent and Bax-mediated intrinsic pathway in HepG2 cells. *Exp. Cell Res.* **2015**, *336*, 308–317. [CrossRef]
44. Lu, J.J.; Meng, L.H.; Cai, Y.J.; Chen, Q.; Tong, L.J.; Lin, L.P.; Ding, J. Dihydroartemisinin induces apoptosis in HL-60 leukemia cells dependent of iron and p38 mitogen-activated protein kinase activation but independent of reactive oxygen species. *Cancer Biol. Ther.* **2008**, *7*, 1017–1023. [CrossRef]
45. Kim, H.P.; Lee, J.Y.; Jeong, J.K.; Bae, S.W.; Lee, H.K.; Jo, I. Nongenomic Stimulation of Nitric Oxide Release by Estrogen Is Mediated by Estrogen Receptor α Localized in Caveolae. *Biochem. Biophys. Res. Commun.* **1999**, *263*, 257–262. [CrossRef]
46. Kim, J.-Y.; Choi, J.-Y.; Lee, H.-J.; Byun, C.J.; Park, J.-H.; Park, J.H.; Cho, H.-S.; Cho, S.-J.; Jo, S.A.; Jo, I. The Green Tea Component (-)-Epigallocatechin-3-Gallate Sensitizes Primary Endothelial Cells to Arsenite-Induced Apoptosis by Decreasing c-Jun N-Terminal Kinase-Mediated Catalase Activity. *PLoS ONE* **2015**, *10*, e0138590. [CrossRef] [PubMed]
47. Tsai, M.H.; Liu, J.F.; Chiang, Y.C.; Hu, S.C.; Hsu, L.F.; Lin, Y.C.; Lin, Z.C.; Lee, H.C.; Chen, M.C.; Huang, C.L.; et al. Arto-carpin, an isoprenyl flavonoid, induces p53-dependent or independent apoptosis via ROS-mediated MAPKs and Akt ac-tivation in non-small cell lung cancer cells. *Oncotarget* **2017**, *8*, 28342–28358. [CrossRef]
48. Cho, D.H.; Choi, Y.J.; Jo, S.A.; Jo, I. Nitric oxide production and regulation of endothelial nitric-oxide synthase phos-phorylation by prolonged treatment with troglitazone: Evidence for involvement of peroxisome proliferator-activated receptor (PPAR) gamma-dependent and PPARgamma-independent signaling pathways. *J. Biol. Chem.* **2004**, *279*, 2499–2506. [PubMed]
49. Kim, H.Y.; Yu, Y.; Oh, S.-Y.; Wang, K.-K.; Kim, Y.-R.; Jung, S.-C.; Kim, H.S.; Jo, I. Far-Infrared Irradiation Inhibits Adipogenic Differentiation and Stimulates Osteogenic Differentiation of Human Tonsil-Derived Mesenchymal Stem Cells: Role of Protein Phosphatase 2B. *Cell. Physiol. Biochem.* **2019**, *52*, 240–253. [CrossRef] [PubMed]

toxins MDPI

Article

Zearalenone Induces Apoptosis and Cytoprotective Autophagy in Chicken Granulosa Cells by PI3K-AKT-mTOR and MAPK Signaling Pathways

Yifeng Zhu [1,†], Heng Wang [2,†], Jianping Wang [1,†], Shunshun Han [2,†], Yao Zhang [2], Menggen Ma [3], Qing Zhu [2], Keying Zhang [1] and Huadong Yin [2,*]

1 Institute of Animal Nutrition, Key Laboratory for Animal Disease-Resistance Nutrition of China, Ministry of Education, Sichuan Agricultural University, Chengdu 611130, China; zhuyifeng@stu.sicau.edu.cn (Y.Z.); wangjianping@sicau.edu.cn (J.W.); zhangkeying@sicau.edu.cn (K.Z.)
2 Farm Animal Genetic Resources Exploration and Innovation Key Laboratory of Sichuan Province, Sichuan Agricultural University, Chengdu 611130, China; wangheng@stu.sicau.edu.cn (H.W.); hanshunshun@stu.sicau.edu.cn (S.H.); zhangyao@sicau.edu.cn (Y.Z.); zhuqing@sicau.edu.cn (Q.Z.)
3 College of Resources, Sichuan Agricultural University, Chengdu 611130, China; mgen@sicau.edu.cn
* Correspondence: yinhuadong@sicau.edu.cn; Tel.: +86-018980787717
† These authors contributed equally to this work.

Abstract: Zearalenone (ZEA) is a nonsteroidal estrogenic mycotoxin found in several food commodities worldwide. ZEA causes reproductive disorders, genotoxicity, and testicular toxicity in animals. However, little is known about the functions of apoptosis and autophagy after exposure to ZEA in granulosa cells. This study investigated the effects of ZEA on chicken granulosa cells. The results show that ZEA at different doses significantly inhibited the growth of chicken granulosa cells by inducing apoptosis. ZEA treatment up-regulated Bax and downregulated Bcl-2 expression, promoted cytochrome c release into the cytosol, and triggered mitochondria-mediated apoptosis. Consequently, caspase-9 and downstream effector caspase-3 were activated, resulting in chicken granulosa cells apoptosis. ZEA treatment also upregulated LC3-II and Beclin-1 expression, suggesting that ZEA induced a high level of autophagy. Pretreatment with chloroquine (an autophagy inhibitor) and rapamycin (an autophagy inducer) increased and decreased the rate of apoptosis, respectively, in contrast with other ZEA-treated groups. Autophagy delayed apoptosis in the ZEA-treated cells. Therefore, autophagy may prevent cells from undergoing apoptosis by reducing ZEA-induced cytotoxicity. In addition, our results further show that the autophagy was stimulated by ZEA through PI3K-AKT-mTOR and MAPK signaling pathways in chicken granulosa cells.

Keywords: zearalenone; apoptosis; autophagy; granulosa cells; PI3K-AKT-mTOR; MAPK; chicken

Key Contribution: In this study, we found that ZEA could protect cell apoptosis by activating autophagy, and it was found that ZEA could regulate autophagy and apoptosis through PI3K-Akt-mTOR and MAPK signaling pathways.

Citation: Zhu, Y.; Wang, H.; Wang, J.; Han, S.; Zhang, Y.; Ma, M.; Zhu, Q.; Zhang, K.; Yin, H. Zearalenone Induces Apoptosis and Cytoprotective Autophagy in Chicken Granulosa Cells by PI3K-AKT-mTOR and MAPK Signaling Pathways. *Toxins* **2021**, *13*, 199. https://doi.org/10.3390/toxins13030199

Received: 2 February 2021
Accepted: 8 March 2021
Published: 10 March 2021

1. Introduction

Zearalenone (ZEA) is an estrogen-like non-steroidal mycotoxin produced by a variety of fusarium fungi. It is often found in grain crops and animal feed as a pollutant, causing serious harm to animal husbandry [1,2]. ZEA has very stable chemical properties and does not deactivate during feed processing. Numerous studies have confirmed the harmful effects of exposure to ZEA and its metabolites on animals and humans, resulting in a variety of diseases and significant economic losses [3,4]. Many reports have shown that exposure to ZEA causes early puberty in children, endometrial adenocarcinoma, female breast cancer, and reduced testicular germ cells [5,6]. Meanwhile, in livestock production, low and high concentrations of ZEA have all been shown to have adverse effects, leading to hyperestrogenemia, abortion, and reproductive failure [7–9].

ZEA has high cytotoxicity, there have been plenty of experiments demonstrating that ZEA can induce the apoptosis of a variety of cells, such as TM4 Sertoli cells, the pig oocytes, male reproductive cells, testicular germ cells, MCF-7 cells, liver cells, and thymus lymphocytes, which suggests that ZEA can induce cell toxicity and apoptosis in many types of cells bearing high proliferation activity [1–3].

Follicle granulosa cells are the largest group of cells in the follicle. They play an important role in follicle development and are one of the important signs of follicle development. Follicular granulosa cells support oocyte development by providing necessary nutrients [4]. Granulosa cells are also involved in maintaining oocyte meiosis retardation, inhibiting oocyte transcriptional activity and inducing oocyte meiosis and cytoplasmic maturation [5]. In addition, granulosa cells are also related to the local micro-environmental control system of the ovary and apoptosis of granulosa cells. Apoptosis may lead to follicular artemia [6]. Recent studies have shown that relatively high concentrations of ZEA can induce apoptosis and necrosis of porcine granulosa cells, which may result in the interruption of steroids [7]. However, no effects of ZEA on chicken granulosa cells have been reported. Recent studies have indicated that ZEA induces apoptosis and cytoprotective autophagy in primary Leydig cells [2]. Therefore, our study aimed to investigate the effects of ZEA at different concentrations in apoptosis and autophagy of chicken granulosa cells, providing experimental evidence of the potential molecular mechanisms underlying ZEA-induced chicken granulosa cells.

2. Results

2.1. Analysis of the Cell Viability

To investigate whether ZEA had an adverse effect on the viability of chicken granulosa cells, we cultured chicken granulosa cells. When grown to approximately 90%, cells with different concentrations of ZEA for 24 h were treated and the cell viability was assessed by the 3-(4,5)-dimethylthiahiazo (MTT) Assay Kit. As shown in Figure 1, the viability of granulosa cells decreased gradually as the concentration of ZEA increased. The survival rate of ZEA treated cells with the same or less than 20 μM was over 68.3%. The survival rate was 52.14% in the 40 μM group, which was close to the 50% inhibitive concentration (IC50). Thus, we choose the 1, 5, 10, 20, and 40 μM ZEA as the treatment concentrations.

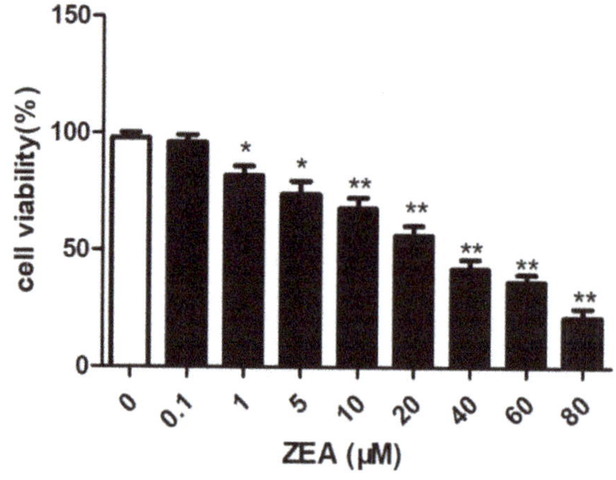

Figure 1. Analysis of the cell viability. The MTT kit was used to analyze the cell viability of chicken granulosa cells. The results are presented as mean ± SD. * $p < 0.05$, ** $p < 0.01$ versus the control group.

2.2. ZEA Induces Apoptosis in Chicken Granulosa Cells

To examine ZEA-induced apoptosis of chicken granulosa cells, Annexin V/PI dual staining was used for cells treated with different concentrations of ZEA for 24 h. The apoptotic rate significantly increased from 8.7% in the control group to 34.8% in the ZEA-treated group (Figure 2a). Then, the cells apoptosis morphological changes were determined using a terminal deoxynucleotidyl transferase-mediated dUTP-biotin nick end labeling (TUNEL) assay. TUNEL assay revealed that ZEA induced a remarkable increase in the TUNEL-positive cells in chicken granulosa cells treated with ZEA in a dose-dependent manner compared to the control (Figure 2b). Quantitative PCR (qPCR) showed that the caspase-3, caspase-8, and caspase-9 mRNA expression increased after being treated with ZEA (Figure 2c). In addition, apoptosis-associated factors gene caspase-3 and caspase-9 were detected via a Western blot. The Western blot showed that the protein levels of activated cleaved caspase-3 and caspase-9 significantly increased in a dose-dependent manner (Figure 2d). These results suggest that ZEA treatment induced apoptosis in chicken granulosa cells in a dose-dependent manner.

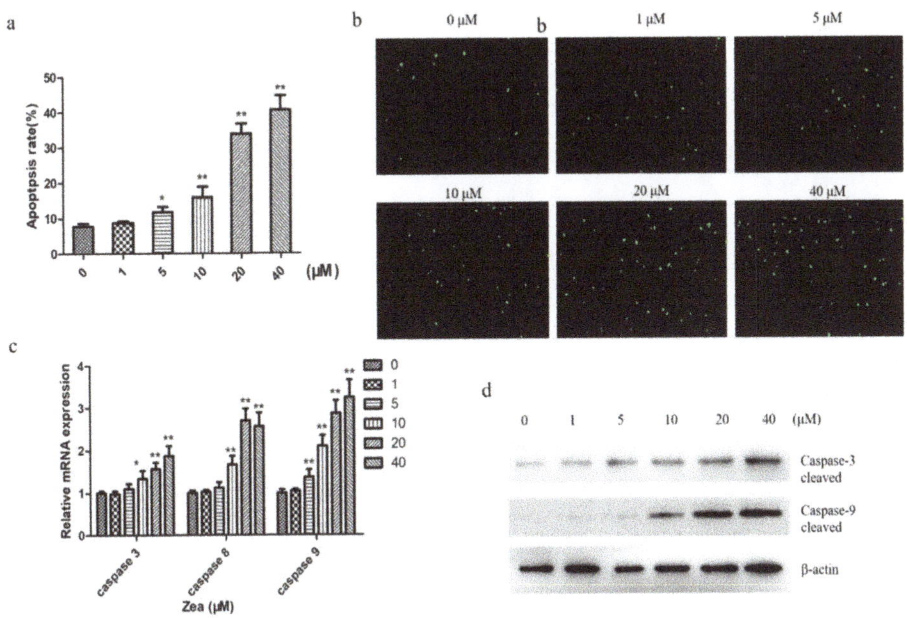

Figure 2. ZEA induces apoptosis in chicken granulosa cells. (**a**) Flow cytometry was used to detect the apoptosis of chicken granulosa cells treated with ZEA. (**b**) Chicken granulosa cells treated with different concentrations of ZEA were detected by TUNEL staining. (**c**) The mRNA expression levels of caspase-3, caspase-8 and caspase-9 in ZEA-treated chicken granulosa cells were detected by qPCR. (**d**) Western blot was used to detect the protein levels of cleaved caspase 3 and caspase 9. The results are presented as mean ± SD. * $p < 0.05$, ** $p < 0.01$ versus the control group.

2.3. The Mitochondrial Apoptotic Pathway was Activated by ZEA

To find out the potential cause of ZEA-induced apoptosis, we next focused on whether it is caused by the mitochondrial apoptotic pathway. We first explored whether ZEA has an effect on the intracellular ROS levels of chicken granulosa cell. The results show that chicken granulosa cells treated with ZEA had significantly enhanced ROS in a dose-dependent manner compared to control (Figure 3a,b). We evaluated the expression levels of Bax and Bcl-2 by Western blot to assess whether the mitochondrial pathway is involved in ZEA-induced apoptosis (Figure 3c). In addition, the activity of Sod1, Cat, Gpx1, and GSH was downregulated with increasing ZEA concentration (Figure 3d). Then, we explored the mitochondrial release of cytochrome c (Cyt c) during ZEA-induced apoptosis. Western blot analysis showed that the level of mitochondrial Cyt c decreased in a dose-dependent manner and that the level of cytosolic Cyt c concomitantly increased (Figure 3e). Collectively, our results show that oxidative stress induced by ZEA triggers apoptosis through the mitochondrial pathway.

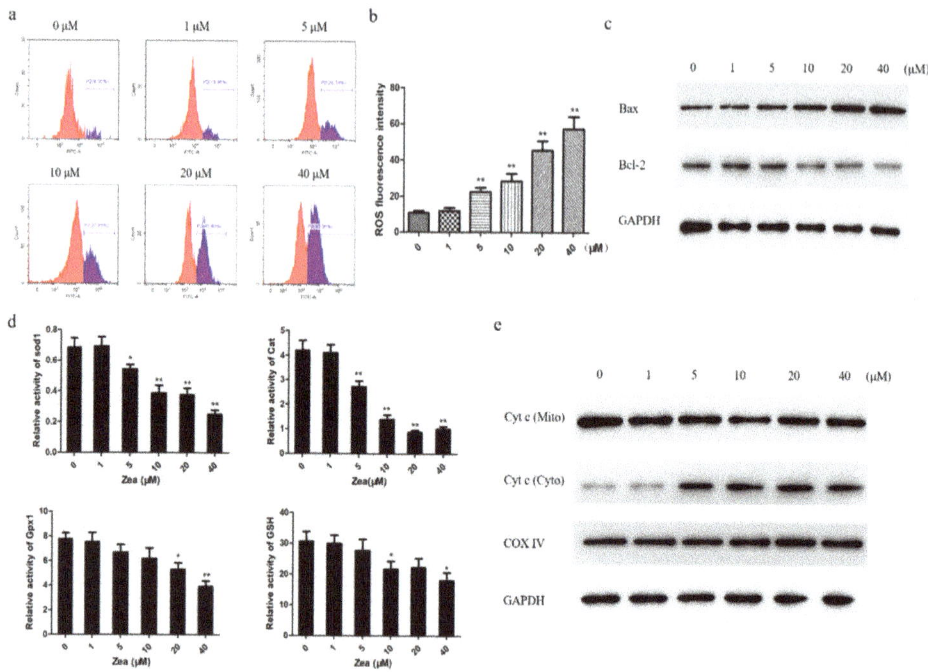

Figure 3. The mitochondrial apoptotic pathway was activated by ZEA. (**a**) Flow cytometry detected intracellular ROS with different concentrations of ZEA treated. (**b**) The bar chart indicated the average intensity of reactive oxygen species. (**c**) The protein expression level of Bax and Bcl-2 in chicken granulosa cells with ZEA treated. (**d**) The enzymatic activities of SOD1, CAT, GPX1, and GSH in chicken granulosa cells. (**e**) The expression level of cytosolic and mitochondrial fractions cytochrome c protein. Cytochrome c oxidase IV (COX IV) and glyceraldehyde-3phosphate dehydrogenase (GAPDH) were used as internal controls for the mitochondrial and cytosolic fractions, respectively. The results are presented as mean ± SD. * $p < 0.05$, ** $p < 0.01$ versus the control group.

2.4. ZEA Induces Autophagy and Delays Apoptosis in Chicken Granulosa Cells

It has been reported that ROS are essential for autophagy and specifically regulating the activity of autophagy genes, and we further explored the influence of ZEA in autophagy. The data from Western blotting showed that the expressions of LC3-II and Beclin-1 were increased significantly in a concentration-dependent manner after the 24 h ZEA treatment. The expression of the P62 protein was significantly decreased after the treatment with ZEA compared to the control group (Figure 4a). As shown in the result, the ZEA-treated groups showed a remarkable increase in LC3-positive puncta as compared to the control by confocal immunohistochemistry in chicken granulosa cells (Figure 4b). Finally, these data suggested that ZEA can trigger the autophagy in chicken granulosa cells.

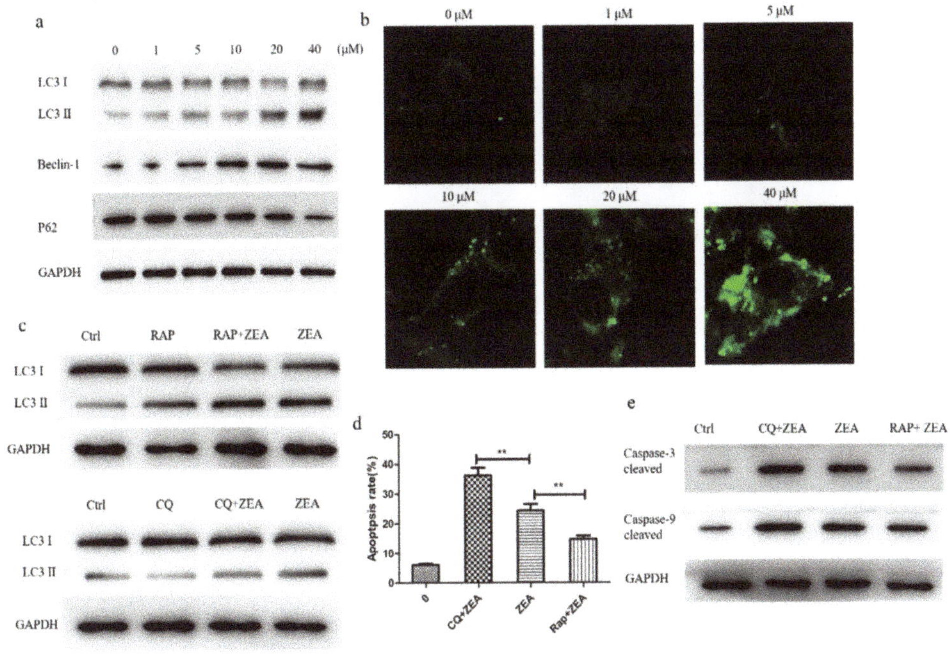

Figure 4. ZEA triggers autophagy in chicken granulosa cells. (**a**) The protein expression levels of LC3, beclin-1, and P62. (**b**) After treated with ZEA for 24 h, LC3 puncta were observed under the fluorescence microscopy. (**c**) LC3 expression after incubated with rapamycin (RAP) and chloroquine (CQ). (**d**) The percentages of apoptotic cells. (**e**) The protein expression level of cleaved caspase-3 and caspase-9 after being incubated with RAP and CQ. The results are presented as mean ± SD. * $p < 0.05$, ** $p < 0.01$ versus the control group.

To examine the effect of ZEA-induced autophagy on apoptosis, the specific autophagic inhibitor chloroquine (CQ) and inducer rapamycin (RAP) were introduced to the ZEA-treated granulosa cells to elucidate the relationship between apoptosis and autophagy in ZEA induced toxicity (Figure 4c). The results show that the apoptotic rate significantly increased after the cotreatment with 20 μM ZEA and CQ compared with that after the ZEA treatment alone and then reversed, from 23% to 48%. By contrast, the apoptotic rate decreased by 15% after the cotreatment with 20 μM ZEA and RAP (Figure 4d). Western blot also found that RAP treatment significantly reduced the protein expression of caspase-3 and caspase-9, and the protein level significantly increased after the addition of autophagy inhibitor CQ (Figure 4e). This result suggests that autophagy hinders apoptosis in ZEA-treated chicken granulosa cells.

2.5. *ZEA Inhibits PI3K/AKT/mTOR and MAPK Signaling Pathway in Chicken Granulosa Cells*

To investigate the mechanisms of ZEA-induced autophagy, we performed RNA sequencing analysis on control and ZEA-treated chicken granulosa cells. We found many autophagy- and apoptosis-related genes, such as caspase-3, Bcl-6, ULK1, and ATG8 (Figure 5a). We enriched 24 signaling pathways through KEGG analysis, and MAPK and mTOR were shown to be associated with autophagy and apoptosis (Figure 5b). Then, we monitored the effect of ZEA on the PI3K/AKT/mTOR and MAPK signaling pathways. The results from Western blot suggest that compared with the control group, the ratios of p-PI3K/PI3K, p-AKT/AKT, and p-mTOR/mTOR were significantly decreased after the treatment with ZEA (Figure 5c,d). Similarly, the ratios of MAPK family proteins p-ERK1/2/ERK1/2, p-JNK1/2/JNK1/2 and p-P-38/P38 were also significantly decreased in a concentration dependent manner (Figure 5e,f). These data indicated that the PI3K-Akt-mTOR and ERK signaling pathways were involved in the process of ZEA-induced autophagy.

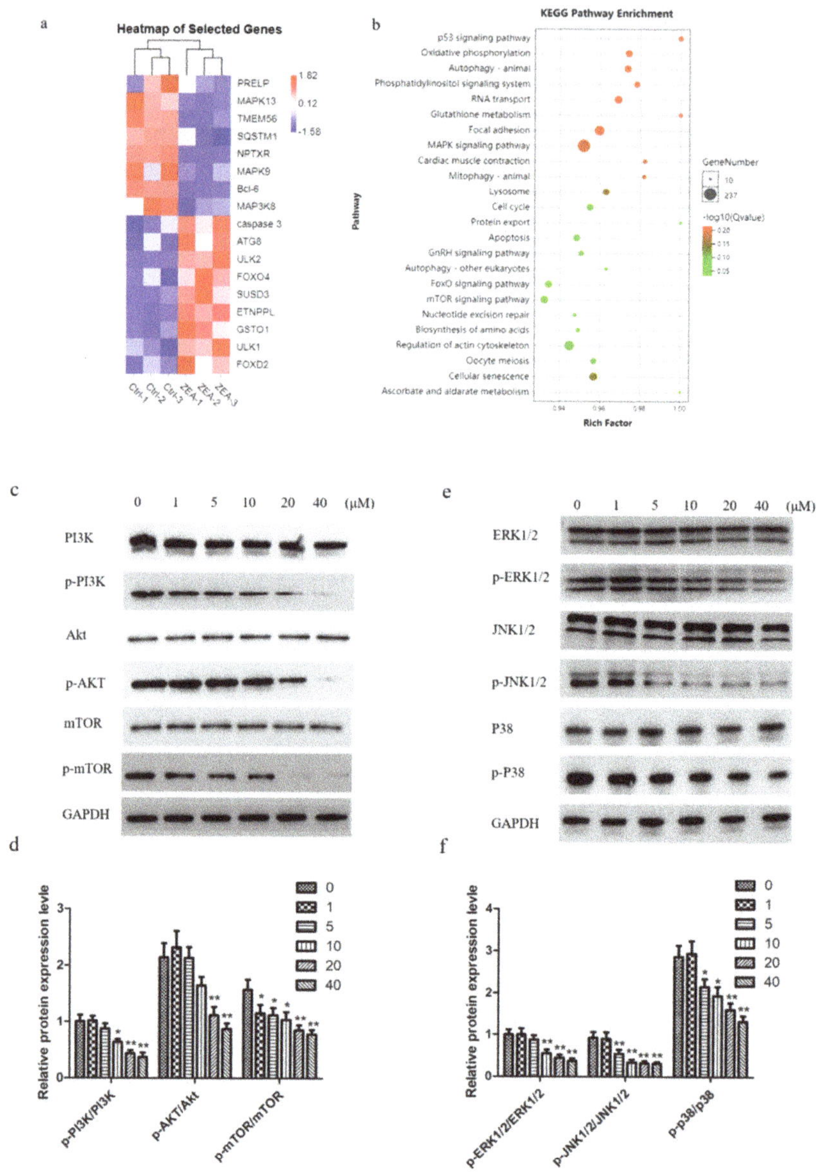

Figure 5. The effects of ZEA on the MAPK family proteins and PI3K-AKT-mTOR pathway. (**a**) Heat map of changes in expression levels of selected genes 24 h after ZEA treatment (20μM). (**b**) Functional enrichment pathways of chicken granulosa cells treated with ZEA. (**c**,**d**) Western blot determined the expression of p-PI3K, PI3K, p-AKT, AKT, p-mTOR, and mTOR. (**e**,**f**) The expression of MAPK family proteins detected by Western blot. The results are presented as mean ± SD. * $p < 0.05$, ** $p < 0.01$ versus the control group.

3. Discussion

The toxicity of ZEA has attracted widespread attention due to the estrogenic effects of its metabolites. Over the past decade, both in vivo and in vitro studies have shown that ZEA has harmful effects on the reproductive systems of humans and animals [8–10]. Although studies have observed ZEA inducing autophagy and apoptosis in mammalian granulosa cells, the molecular mechanism of ZEA cytotoxicity remains unclear [3,11]. Therefore, the purpose of this study was to investigate the molecular mechanisms of apoptosis and maize autophagy.

Granulosa cells, as the main unit of follicles, are the key to ensuring oocyte maturation and maintaining normal hormone levels in animals. Their main functions include the production of sexual steroids and numerous growth factors that are thought to interact with oocytes during development [12]. In our experiment, MTT showed that ZEA inhibited the proliferation of chicken granulosa cells in a concentration-dependent manner and showed cytotoxic reactions under the conditions of 24 h exposure, and similar results were obtained in other relevant studies [13,14]. Apoptosis is a kind of programmed cell death, which plays an important role in embryogenesis, metamorphosis, and cell homeostasis [15]. We found that ZEA induced significant apoptosis of chicken granulosa cells in a dose-dependent manner. ZEA induces apoptosis in many cell types, including male rat germ cells, human leukemia cells, CTLL-2 cells, porcine oocytes, mouse support cells, and so on, which is consistent with previous reports [16–20]. Caspase plays a vital role in apoptosis by cleaving large amounts of proteins [21]. Caspase cascade activation triggers apoptosis in ZEA-induced cells [13,22]. Inhibiting the activity of caspases can significantly reduce the apoptotic effect of ZEA. These results indicate that caspase cascade activation was involved in the ZEA-induced apoptosis of chicken granulosa cells. These results are consistent with our results. We also found that ZEA induced decreased cell viability and induced the increase in apoptosis-related proteins, leading to cell apoptosis. This may be the reason for the decrease in laying rate and the death of chickens caused by mildewed feed in the poultry rearing process.

As is known, the mitochondrial pathway plays a central role in the apoptosis pathway, which determines the survival or death [23]. In this study, we found that ZEA leads to the loss of mitochondrial transmembrane potential and the increase in ROS in chicken granulosa cells in a dose-dependent manner. In addition, in the ZEA-treated group, the expression ratio of Bax/Bcl-2 protein increased, while Cyt c was released from the mitochondria. Loss of transmembrane potential is regarded as a major determinant of cell involvement at death [24]. ROS has been reported to directly activate mitochondrial permeability transformation, leading to mitochondrial transmembrane potential loss and Cyt C release [25]. In cells, the mitochondrial outer membrane's permeability and Cyt C translocated from the mitochondrial intermembrane compartment into the cytoplasm through Bcl-2 family proteins are two major factors upon which mitochondrial function depends [26]. Cyt c and Apaf-1 activate caspase-9, which activates executioner caspase-3. The Bax/Bcl-2 pathway participates in cell apoptosis, and its relative expression level decides the final fate of cells [27]. ZEA has been reported to induce apoptosis in cells through ROS-mediated mitochondrial pathways, such as RAW264.7 cells, swine IPE-J2 cells, and swine granulosa cells. Our results are consistent with those reported above [13,28,29].

The intracellular degradation system and the dynamic circulation system trigger autophagy, and the bi-membrane vacuole transport and transport of cytoplasm in lysase for degradation [30]. Under normal physiological conditions, cells maintain low levels of autophagy. However, under oxidative stress, endoplasmic reticulum stress, nutritional restriction, and other stress conditions, cells can trigger autophagy and survive [31]. Several studies showed that a variety of cells treated with ZEA can activate autophagy and oxidative stress [3,11,32]. Our experimental data agree with these results, such as increasing beclin 1 and LC3 II protein expression can induce p62 degradation, which suggests that ZEA can promote autophagy flux. Beclin-1 is an autophagic protein that can regulate the initiation of autophagy and the fusion of autophagy-lysosome, interacting with Atg14L and the Rubicon protein complex. On the autophagosome membrane, LC3-I is transformed into LC3-II, and a circular structure is formed in the cytoplasm. Changes in the synthesis and processing of LC3 during autophagy make it a specific biomarker of autophagy. Meanwhile, the recognition and transport of ubiquitination proteins to autophagy for degradation is the function of P62 as a selective autophagy receptor.

After proving that ZEA induces apoptosis and autophagy, we investigated the role of autophagy in apoptosis. Inhibition of autophagy by CQ improves the apoptosis rate, leading to ZEA-induced apoptosis of chicken granulosa cells, compared with RAP-induced autophagy reducing the apoptosis rate. Our study found that autophagy promoted survival because inhibition of autophagy aggravated ZEA-induced cytotoxicity and apoptosis. ZEA has been reported to induce apoptosis of primary mesenchymal cells and protective autophagy. This suggests that autophagy can alleviate ZEA-induced apoptosis to some extent. Autophagy is closely involved in many physiological pathways, such as apoptosis. Like autophagy, apoptosis plays a decisive role in cell development and growth [33]. In fact, autophagy and apoptosis use common proteins, and it is widely accepted that increased autophagy is a protective mechanism against apoptotic cell death. In the process of the disease, cancer cells increase the resistance to apoptosis induced by chemotherapy by increasing the flow of autophagic and removing damaged organs and proteins [34]. Therefore, the increased sensitivity to apoptosis induced by ZEA treatment may also prevent apoptosis through increased autophagy.

The PI3K/AKT/mTOR and MAPK pathway are two well-known pathways involved in the regulation of autophagy [35]. MAPK signaling pathways interfere with autophagy in a variety of types of cells [36]. The mTOR protein is the main regulatory factor involved in the induction of autophagy, whose up-regulation can decrease the level of autophagy, whereas its down-regulation can trigger the phosphorylation of mTOR and finally increase the level of autophagy. It is the consensus that PI3K activates AKT and in doing so leads to the phosphorylation and activation of mTOR [37]. ZEA has been shown to inhibit PI3K/AKT/mTOR pathway. In our study, we found that ZEA inhibited PI3K, AKT, mTOR, p70S6K, and ERK, while it activated JNK and p38. Studies have shown that ZEA can stimulate autophagy through PI3K/AKT/mTOR and MAPK signaling pathways in different cell types. Our results are consistent with the above reports [1,28,38]. Moreover, the current study suggests that ERK1/2 and PI3K/AKT/mTOR signaling pathways were concerned with the autophagy activation.

4. Conclusions

ZEA treatment of primary chicken granulosa cells inhibited cell viability. ZEA treatment activates the mitochondrial apoptosis pathway through Bcl-2 family proteins and eventually leads to cell apoptosis. ZEA also increased the level of autophagy, delayed apoptosis, and activated PI3K/ Akt /mTOR and AMPK signaling pathways. In summary, this study suggests that ZEA regulates the autophagy and apoptosis of chicken granulosa cells through the PI3K/AKT/mTOR and AMPK pathways.

5. Materials and Methods

5.1. Animals

Sexually mature, egg-laying Roman pink hens (25–35 weeks of age) were used in this study (Chengdu, Sichuan, China). Hens were raised in the institute "vivarium" under controlled conditions with a photoperiod of 13 h light/11 h dark and were fed 100 g/lb. of commercial layer feed (45% carbohydrate, 25% protein, 20% fat, 10% calcium and other minerals, US Purina Chow) (Zhengda, Neimeng, China) and water per day. All experimental animals were executed by decapitation (in accordance with the Institute's Bioethical Committee regulations). The follicles in the pre-graded development stage (3–10, 115 mm) were collected in normal saline (0.9% NaCl) containing 1% penicillin streptomycin when the animals were used for laying eggs every day. The housing, feeding, and all experimental programs used in this study strictly followed the guidelines of the Animal

Welfare Committee of the College of Agriculture, Sichuan Agricultural University, whose approval number is 20191078632 (2019-03-05).

5.2. Granulosa Cells Culture

According to the previous study by Gilbert et al. [39], culturing of primary granulosa cells was performed as previous described [40]. Briefly, the layer of follicular granulosa cells was carefully removed from the follicular wall in a sterile manner. After five washes in PBS, follicular granulosa cells were incubated in DMEM/F12 (Sigma, St. Louis, MO, USA) with 7% fetal bovine serum (FBS, Gibco Invitrogen, Waltham, MA, USA) and 1% primocin (InvivoGen, San Diego, CA, USA). Then, we made the monolayer of granulosa cells in a protease solution containing 0.1% collagenaseII (Worthington, Lakewood, NJ, USA) and 0.05% trypsin (Sigma, St. Louis, MO, USA) in PBS for digestion with gently agitation for 15 min at 37 °C, and then the solution was centrifuged at 2000 rpm for 5 min. After the supernatant was discarded and the cells were re-suspended, cells were filtered through a 40 μM sieving nylon mesh. Cells were incubated for 2 h and incubated with 250 μL fresh media for 24 h at 37 °C in the condition of 5% CO2.

5.3. Intracellular Reactive Oxygen Species (ROS)

ZEA was purchased from Sigma-Aldrich (St. Louis, MO, USA), and was then diluted in DMSO. Six different concentrations of ZEA (0, 1, 5, 10, 20, and 40 μM)-treated cells were added to 10mM DCFH-DA for 20 min incubation at 37 °C. After removing the extracellular DCFH-DA by serum-free medium washing, flow cytometry was performed to detect granulosa cells' intracellular ROS levels with or without ZEA treatment.

5.4. Cell Viability Assay

The MTT assay was used to detect the effects on ZEA-treated cell viability. Granulosa cells were cultured in 96-well plates. When reaching a density of 5×10^4 cells per well, cells were treated with different concentrations of ZEA (2.5, 5, 10, and 20 μm) for 24 h. The cells were treated with 10 μL 10 mg/mL MTT, and the absorbance was determined at 490 nm. The results were expressed as a percentage of the control group, which was randomly assigned to a 100% viability.

5.5. Flow Cytometry Apoptosis

After treatment, the cells were then stained dark at room temperature for 15 min in a 100 μL binding buffer with 5 μL Annexin V-FITC and 5 μL propidium iodide as instructed. Flow cytometer was performed to determine the fluorescent signal (FACS Calibur; Becton–Dickinson, Franklin Lakes, NJ, USA).

5.6. Western Blotting Analysis.

The harvested cells were lysed by ultrasonication. The proteins were transferred to PVDF membranes (Millipore Corporation, Billerica, MA, USA) after gel electrophoresis (SDS-PAGE). The membranes were incubated at room temperature with 5% defatted milk powder for 1 h and then probed with the indicated primary antibodies: cleaved caspase-3, cleaved caspase-9, Bax, Bcl-2, LC3, beclin-1, P62, PI3K, p-PI3K, AKT, p-AKT, mTOR, p-mTOR, ERK1/2, p-ERK1/2, JNK1/2, p-JNK1/2, P38, p-P38 (all from Zen Bioscience, Chengdu, China) at 4 °C overnight. Next, the following secondary antibody was employed and incubated 2 h at room temperature. The signal was developed by the ECL detection system, and the relative expression of proteins was analyzed using the Quantity One software.

5.7. Immunofluorescence Assay

The cells were treated with ZEA for 24 h and then 0.5% Triton X-100 was used for fixation in 4% paraformaldehyde for 30 min at room temperature, and 5% BSA was used for sealing. LC3 antibody (1:2000) was incubated in the blocking solution at 4 °C overnight, and then FITC-conjugated anti-mouse secondary IgG (1:1000) was incubated at room temperature for 1 h. The samples were examined with a fluorescence microscope (Leica 2500; Leica Corpo-ration, Germany).

5.8. TUNEL Assay

Chicken granulosa cells were washed with PBS twice and fixed in 4% paraformaldehyde for 15 min. To detect DNA fragments, TUNEL analysis was performed using in situ cell death detection kits (Solarbio, Beijing, China) according to the manufacturer's instructions. After Tunel-positive cells were mounted, they were then observed under a fluorescence microscope (Olympus, Tokyo, Japan).

5.9. RNA Isolation and qRT-PCR

Total RNA was extracted with TRIzol reagent (Takara, Tokyo, Japan) according to the manufacturer's instructions. Approximately 2 μg RNA was reverse transcribed using the Takara PrimeScript RT reagent kit (Takara) according to the manufacturer's instructions. The qRT-PCR was performed essentially as described previously [27]. Statistical analysis of the RT-PCR results was performed by determining mean threshold cycle (ΔCt) values for the expression of standardized genes. The primers used are shown as follows: caspase-3 forward: TGGCCCTCTTGAACTGAAAG; caspase-3 reverse: TCCACTGTCT-GCTTCAATACC; caspase8 forward: CCCTGAAGACAGTGCCATTT; caspase-8 reverse: GGGTCGGCTGGTCATTTTAT; caspase-9 forward: TCCCGGGCTGTTTCAACTT; caspase-9 reverse: CCTCATCTT-GCAGCTTGTGC; GAPDH forward: TCCTCCACCTTTGATGCG; GAPDH reverse: GTGCCTGGCTCACTCCTT.

5.10. Autophagy Analysis

Chicken granulosa cells were cultured in 24-well plates. Cells were treated with rapamycin (5μM, Sigma) for 6h to induce autophagy. To block autophagy, cells were treated with chloroquine (CQ) (10 μM, Sigma) for 6 h.

5.11. Antioxidative Enzymes Detection

The activities of superoxide dismutase 1 (SOD1), glutathione peroxidase (GPX1), catalase (CAT), and glutathione (GSH) were measured by the commercially available kit (Jiancheng, Nanjing) according to the manufacturer's instructions.

5.12. RNA-seq

The total RNA was extracted from control and ZEA-treated cells using previously described materials and methods. The cDNA library construction, sequencing, and transcriptome data analysis were conducted by Guangzhou Gidio Biotechnology Co., Ltd (Giddo, Guangdong, China).

5.13. Statistical Analysis

The results were expressed as mean ± standard deviation (SD). Non-parametric one-way ANOVA with SPSS was used to compare the statistical data between the groups, and $p < 0.05$ was considered statistically significant. Each experiment should be done at least three times.

Author Contributions: Conceptualization, Y.Z., H.W., J.W., S.H., and H.Y.; Data curation, Y.Z., H.W., J.W., and S.H; Formal analysis, Y.Z., H.W., S.H., and M.M.; Funding acquisition, H.Y.; Investigation, Y.Z., J.W., and S.H.; Project administration, Y.Z. (Yao Zhang); Resources, Q.Z.; Supervision, K.Z.; Writing—original draft, Y.Z., H.W., J.W., and S.H.; Writing—review and editing, H.Y. All authors have read and agreed to the published version of the manuscript.

Funding: This work was financially supported by Sichuan Science and Technology Program (2020NZZJ0 and 2019NZZJ0011), and Chengdu Science and Technology Program (2019-YF05-02309-SN).

Institutional Review Board Statement: The Animal Ethics Committee of Sichuan Agricultural University approved all experimental procedures (approval number: 201910201201). Relevant guidelines and regulations were followed for all methods.

Informed Consent Statement: Informed consent was obtained from all subjects involved in the study.

Data Availability Statement: Publicly available datasets were analyzed in this study. the data used in this study have been deposited in the National Center for Biotechnology Information Sequence Red Archive (SRA) under the accession code NCBI accession number: GSE151450. All other data are available within the Article or available from the authors upon request.

Acknowledgments: We thank MDPI for the revision of the manuscript, and we thank the Sichuan Science and Technology Program (2020 NZZJ021 and 2019 NZZJ0011) and the Chengdu Science and Technology Program (2019-YF05-02309SN) for their support.

Conflicts of Interest: The authors declare no conflict of interest.

References

1. Zheng, W.; Wang, B.; Si, M.; Zou, H.; Song, R.; Gu, J.; Yuan, Y.; Liu, X.; Zhu, G.; Bai, J. Zearalenone altered the cytoskeletal structure via ER stress- autophagy- oxidative stress pathway in mouse TM4 Sertoli cells. *Sci. Rep.* **2018**, *8*, 3320. [CrossRef]
2. Wang, Y.; Zheng, W.; Bian, X.; Yuan, Y.; Gu, J.; Liu, X.; Liu, Z.; Bian, J. Zearalenone induces apoptosis and cytoprotective autophagy in primary Leydig cells. *Toxicol. Lett.* **2014**, *226*, 182–191. [CrossRef]
3. Han, J.; Wang, T.; Fu, L.; Shi, L.Y.; Zhu, C.C.; Liu, J.; Zhang, Y.; Cui, X.S.; Kim, N.H.; Sun, S.C. Altered oxidative stress, apoptosis/autophagy, and epigenetic modifications in Zearalenone-treated porcine oocytes. *Toxicol. Res.* **2015**, *4*, 1184–1194. [CrossRef]
4. Eppig, J.J. Intercommunication between mammalian oocytes and companion somatic cells. *Bioessays New Rev. Mol. Cell. Dev. Biol.* **1991**, *13*, 569–574. [CrossRef]
5. Eppig, J.J. Maintenance of meiotic arrest and the induction of oocyte maturation in mouse oocyte-granulosa cell complexes developed in vitro from preantral follicles. *Biol. Reprod.* **1991**, *45*, 824–830. [CrossRef]
6. Yuan, S.Y.; Hong, S.S.; Zheng, B.H.; Wei, L.I.; Ming, J.L.; Jing, H.T. Apoptosis in Granulosa cells during follicular atresia: relationship with steroids and insulin-like growth factors. *Cell Res.* **2004**, *14*, 341–346.
7. Kolesarova, A.; Capcarova, M.; Medvedova, M.; Sirotkin, A.V.; Kovacik, J. In vitro assessment of iron effect on porcine ovarian granulosa cells: Secretory activity, markers of proliferation and apoptosis. *Physiol. Res.* **2011**, *60*, 503–510. [CrossRef]
8. EFSA Panel on Contaminants in the Food Chain. Scientific Opinion on the risks for animal and public health related to the presence of phomopsins in feed and food. *Efsa J.* **2019**, *17*, e05860.
9. Streit, E.; Schatzmayr, G.; Tassis, P.; Tzika, E.; Marin, D.; Taranu, I.; Tabuc, C.; Nicolau, A.; Aprodu, I.; Puel, O. Current Situation of Mycotoxin Contamination and Co-occurrence in Animal Feed—Focus on Europe. *Toxins* **2012**, *4*, 788–809. [CrossRef]
10. Reddy, K.; Salleh, B.; Saad, B.; Abbas, H.K.; Abel, C.A.; Shier, W.T. An overview of mycotoxin contamination in foods and its implications for human health. *Toxin Rev.* **2010**, *29*, 3–26. [CrossRef]
11. Ben, S.I.; Boussabbeh, M.; Pires, D.S.J.; Guilbert, A.; Bacha, H.; Abidessefi, S.; Lemaire, C. SIRT1 protects cardiac cells against apoptosis induced by zearalenone or its metabolites α- and β-zearalenol through an autophagy-dependent pathway. *Toxicol. Appl. Pharmacol.* **2016**, *314*.
12. M'Baye, M.; Hua, G.; Khan, H.A.; Yang, L. RNAi-mediated knockdown of INHBB increases apoptosis and inhibits steroidogenesis in mouse granulosacells. *J. Reprod. Dev.* **2015**, *61*, 391. [CrossRef] [PubMed]
13. Zhu, L.; Yuan, H.; Guo, C.; Lu, Y.; Deng, S.; Yang, Y.; Wei, Q.; Wen, L.; He, Z. Zearalenone induces apoptosis and necrosis in porcine granulosa cells via a caspase-3- and caspase-9-dependent mitochondrial signaling pathway. *J. Cell. Physiol.* **2012**, *227*, 1814–1820. [CrossRef] [PubMed]
14. Qin, X.; Cao, M.; Lai, F.; Yang, F.; Ge, W.; Zhang, X.; Cheng, S.; Sun, X.; Qin, G.; Shen, W. Oxidative Stress Induced by Zearalenone in Porcine Granulosa Cells and Its Rescue by Curcumin In Vitro. *PLoS ONE* **2015**, *10*, e0127551. [CrossRef]
15. Cortiella, J.; Niles, J.; Cantu, A.; Brettler, A.; Pham, A.; Vargas, G.; Winston, S.; Wang, J.; Walls, S.; Nichols, J.E. Influence of acellular natural lung matrix on murine embryonic stem cell differentiation and tissue formation. *Tissue Eng. Part A* **2010**, *16*, 2565–2580. [CrossRef]
16. Kim, I.H.; Son, H.Y.; Cho, S.W.; Ha, C.S.; Kang, B.H. Zearalenone induces male germ cell apoptosis in rats. *Toxicol. Lett.* **2003**, *138*, 185–192. [CrossRef]
17. Banjerdpongchai, R.; Kongtawelert, P.; Khantamat, O.; Srisomsap, C.; Chokchaichamnankit, D.; Subhasitanont, P.; Svasti, J. Mitochondrial and endoplasmic reticulum stress pathways cooperate in zearalenone-induced apoptosis of human leukemic cells. *J. Hematol. Oncol.* **2010**, *3*, 50. [CrossRef] [PubMed]
18. Yamaji, T.; Nakamura, S.; Takematsu, H.; Kawasaki, T.; Kozutsumi, Y. Apoptosis of CTLL-2 cells induced by an immunosuppressant, ISP-I, is caspase-3-like protease-independent. *J. Biochem.* **2001**, *129*, 521–527.

19. Zheng, W.L.; Wang, B.J.; Wang, L.; Shan, Y.P.; Zou, H.; Song, R.L.; Wang, T.; Gu, J.H.; Yuan, Y.; Liu, X.Z. ROS-Mediated Cell Cycle Arrest and Apoptosis Induced by Zearalenone in Mouse Sertoli Cells via ER Stress and the ATP/AMPK Pathway. *Toxins* **2018**, *10*, 24. [CrossRef] [PubMed]
20. Sambuu, R.; Takagi, M.; Namula, Z.; Otoi, T.; Shiga, S.; Rodrigues, D.S.R.; Finkgremmels, J. Effects of exposure to zearalenone on porcine oocytes and sperm during maturation and fertilization in vitro. *J. Reprod. Dev.* **2011**, *57*, 547. [CrossRef] [PubMed]
21. Tomaszewski, J.; Miturski, R.; Semczuk, A.; Kotarski, J.; Jakowicki, J. Tissue zearalenone concentration in normal, hyperplastic and neoplastic human endometrium. *Ginekologia Polska* **1998**, *69*, 363. [PubMed]
22. Ayed-Boussema, I.; Bouaziz, C.; Rjiba, K.; Valenti, K.; Laporte, F.; Bacha, H.; Hassen, W. The mycotoxin Zearalenone induces apoptosis in human hepatocytes (HepG2) via p53-dependent mitochondrial signaling pathway. *Toxicol. In Vitro* **2008**, *22*, 1671–1680. [CrossRef]
23. Skommer, J. Brittain, T. Raychaudhuri, S. Bcl-2 inhibits apoptosis by increasing the time-to-death and intrinsic cell-to-cell variations in the mitochondrial pathway of cell death. *Apoptosis* **2010**, *15*, 1223–1233. [CrossRef]
24. Bras, M.; Queenan, B.; Susin, S.A. Programmed cell death via mitochondria: Different modes of dying. *Biochem. Biokhimiia* **2005**, *70*, 231–239. [CrossRef] [PubMed]
25. Lee, M.G.; Lee, K.T.; Chi, S.G.; Park, J.H. Costunolide induces apoptosis by ROS-mediated mitochondrial permeability transition and cytochrome C release. *Biol. Pharm. Bull.* **2002**, *24*, 303–306. [CrossRef] [PubMed]
26. Estaquier, J.; Vallette, F.; Vayssiere, J.L.; Mignotte, B. The mitochondrial pathways of apoptosis. *J. Beijing Med Univ.* **2002**, *942*, 157.
27. Li, G.Y.; Xie, P.; Li, H.Y.; Hao, L.; Xiong, Q.; Qiu, T. Involment of p53, Bax, and Bcl-2 pathway in microcystins-induced apoptosis in rat testis. *Environ. Toxicol.* **2011**, *26*, 111–117. [CrossRef]
28. Yu, J.Y.; Zheng, Z.H.; Son, Y.O.; Shi, X.; Jang, Y.O.; Lee, J.C. Mycotoxin zearalenone induces AIF- and ROS-mediated cell death through p53- and MAPK-dependent signaling pathways in RAW264.7 macrophages. *Toxicol. In Vitro* **2011**, *25*, 1654–1663. [CrossRef]
29. Fan, W.; Shen, T.; Ding, Q.; Lv, Y.; Li, L.; Huang, K.; Yan, L.; Song, S. Zearalenone induces ROS-mediated mitochondrial damage in porcine IPEC-J2 cells. *J. Biochem. Mol. Toxicol.* **2017**, *31*, e21944. [CrossRef] [PubMed]
30. Klionsky, D.J.; Emr, S.D. Autophagy as a regulated pathway of cellular degradation. *Science* **2000**, *290*, 1717–1721. [CrossRef]
31. Takagi, H.; Matsui, Y.; Sadoshima, J. The role of autophagy in mediating cell survival and death during ischemia and reperfusion in the heart. *Antioxid. Redox Signal.* **2007**, *9*, 1373–1381. [CrossRef]
32. Han, R.; Liang, H.; Qin, Z.H.; Liu, C.Y. Crotoxin induces apoptosis and autophagy in human lung carcinoma cells in vitro via activation of the p38MAPK signaling pathway. *Acta Pharmacol Sin.* **2014**, *35*, 1323–1332. [CrossRef] [PubMed]
33. Moscat, J.; Diaz-Meco, M.T. p62 at the Crossroads of Autophagy, Apoptosis, and Cancer. *Cell* **2009**, *137*, 1001–1004. [CrossRef] [PubMed]
34. Maiuri, M.C.; Zalckvar, E.; Kimchi, A.; Kroemer, G. Self-eating and self-killing: Crosstalk between autophagy and apoptosis. *Nat. Rev. Mol. Cell Biol.* **2007**, *8*, 741–752. [CrossRef] [PubMed]
35. Zhao, R.; Chen, M.; Jiang, Z.; Zhao, F.; Xi, B.; Xu, Z.; Fu, H.; Zhou, K. Platycodin-D Induced Autophagy in Non-Small Cell Lung Cancer Cells via PI3K/Akt/mTOR and MAPK Signaling Pathways. *J. Cancer* **2015**, *6*, 623–631. [CrossRef]
36. Hirota, Y.; Yamashita, S.; Kurihara, Y.; Jin, X.; Aihara, M.; Saigusa, T.; Kang, D.; Kanki, T. Mitophagy is primarily due to alternative autophagy and requires the MAPK1 and MAPK14 signaling pathways. *Autophagy* **2015**, *11*, 332–343. [CrossRef]
37. Herassandoval, D.; Pérezrojas, J.M.; Hernándezdamián, J.; Pedrazachaverri, J. The role of PI3K/AKT/mTOR pathway in the modulation of autophagy and the clearance of protein aggregates in neurodegeneration. *Cell. Signal.* **2014**, *26*, 2694–2701. [CrossRef]
38. So, M.; Sha, S.; Antoniou, M.; Wu, S.; Tanun, K. Mycotoxin Zearalenone induced apoptosis in BEAS-2B cells through generation of ROS and activation of JNK and p38 MAPKs signalling pathways. *BMC Genet.* **2012**, *6*, 231–240.
39. Gilbert, A.B.; Evans, A.J.; Perry, M.M.; Davidson, M.H. A method for separating the granulosa cells, the basal lamina and the theca of the preovulatory ovarian follicle of the domestic fowl (*Gallus domesticus*). *J. Reprod. Fertil.* **1977**, *50*, 179–181. [CrossRef]
40. Ahumada-Solórzano, S.M.; Carranza, M.E.; Pedernera, E.; Rodríguez-Méndez, A.J.; Luna, M.; Arámburo, C. Local expression and distribution of growth hormone and growth hormone receptor in the chicken ovary: Effects of GH on steroidogenesis in cultured follicular granulosa cells. *Gen. Comp. Endocrinol.* **2012**, *175*, 297–310. [CrossRef]

toxins

Article

N-Acetylcysteine Inhibits Patulin-Induced Apoptosis by Affecting ROS-Mediated Oxidative Damage Pathway

Jiayu Liu †, Qi Liu †, Jiahui Han, Jiayu Feng, Tianmin Guo, Zhiman Li, Fenyi Min, Ruyi Jin and Xiaoli Peng *

College of Food Science and Engineering, Northwest A & F University, Yangling 712100, China; liujiayu1002@nwafu.edu.cn (J.L.); liuqi331@126.com (Q.L.); hanjiahui2021888@163.com (J.H.); fengjiayu@nwafu.edu.cn (J.F.); guotianmin666@nwafu.edu.cn (T.G.); lizhiman2001@nwafu.edu.cn (Z.L.); minfenyi@nwafu.edu.cn (F.M.); jinruyi@nwafu.edu.cn (R.J.)

* Correspondence: xiaolipeng@nwafu.edu.cn; Tel.: +86-29-153-1972-9682

† Jiayu Liu and Qi Liu contributed equally to this work.

Abstract: Patulin (PAT) belongs to the family of food-borne mycotoxins. Our previous studies revealed that PAT caused cytotoxicity in human embryonic kidney cells (HEK293). In the present research, we systematically explored the detailed mechanism of ROS production and ROS clearance in PAT-induced HEK293 cell apoptosis. Results showed that PAT treatment (2.5, 5, 7.5, 10 μM) for 10 h could regulate the expression of genes and proteins involved in the mitochondrial respiratory chain complex, resulting in dysfunction of mitochondrial oxidative phosphorylation and induction of ROS overproduction. We further investigated the role of *N*-acetylcysteine (NAC), an ROS scavenger, in promoting the survival of PAT-treated HEK293 cells. NAC improves PAT-induced apoptosis of HEK293 cells by clearing excess ROS, modulating the expression of mitochondrial respiratory chain complex genes and proteins, and maintaining normal mitochondrial function. In addition, NAC protects the activity of antioxidant enzymes, maintains normal GSH content, and relieves oxidative damage. Additionally, 4 mM NAC alleviated 7.5 μM PAT-mediated apoptosis through the caspase pathway in HEK293 cells. In summary, our study demonstrated that ROS is significant in PAT-mediated cytotoxicity, which provides valuable insight into the management of PAT-associated health issues.

Keywords: PAT; apoptosis; ROS; oxidative stress; mitochondrial oxidative phosphorylation

Key Contribution: We clearly conclude that exposure to patulin led to dysfunction of mitochondrial oxidative phosphorylation and induction of reactive oxygen species overproduction; we also found that *N*-acetylcysteine supplementation ameliorated oxidative damage induced by patulin and alleviated patulin-mediated apoptosis via the caspase pathway in human embryonic kidney cells.

Citation: Liu, J.; Liu, Q.; Han, J.; Feng, J.; Guo, T.; Li, Z.; Min, F.; Jin, R.; Peng, X. *N*-Acetylcysteine Inhibits Patulin-Induced Apoptosis by Affecting ROS-Mediated Oxidative Damage Pathway. *Toxins* **2021**, *13*, 595. https://doi.org/10.3390/toxins13090595

Received: 17 June 2021
Accepted: 23 August 2021
Published: 26 August 2021

1. Introduction

Patulin (PAT), a common mycotoxin, is a food contaminant produced by several fungal species such as *Byssochlamys*, *Aspergillus*, and *Penicillium* [1–3]. PAT is usually present in moldy food materials including apples and its derived products, vegetables, cereals, and cheese [4–6]. PAT is a polyketide lactone which reacts strongly with thiol groups [7,8]. Increasing toxicological evidence has suggested that human and livestock exposure to PAT produces several toxic effects, such as nephrotoxicity, hepatotoxicity, neurotoxicity, and gastrointestinal and dermal toxicities [3,9,10]. Several compounds were found to have protective effects on nephrotoxicity caused by PAT [3,11]. Our previous studies revealed that PAT exerts significant cytotoxicity via the production of intracellular ROS in HEK293 cells [12]. Moreover, studies have shown that ROS-mediated oxidative stress plays a major role in PAT-induced cytotoxicity. However, the underlying molecular mechanism of ROS generation in response to PAT exposure is still unclear.

Mitochondria are thought to be the cell's "powerhouse" and play pivotal roles in many metabolic processes, including energy metabolism, oxidative stress, and apoptosis regulation [13–15]. The mitochondrial oxidative phosphorylation (OXPHOS) system involves the complexes of the electron transfer chain (ETC) and ATP synthase complex. In these compounds, complex I is NADH-ubiquinone reductase, complex II is succinate dehydrogenase, complex III is ubiquinol cytochrome c reductase, complex IV is cytochrome c oxidase, and complex V is ATP-synthase [1,16,17]. Cells performing OXPHOS generate the mitochondrial membrane potential, drive 90% of cellular ATP generation [18], and contribute significantly to cellular ROS. Therefore, regulation of OXPHOS is critical to maintain cellular homeostasis.

The research showed that that ROS play a key role in different cellular processes such as gene expression, energy metabolism, and cell signaling [19–21]. In normal physiological conditions, the antioxidant enzyme system usually regulates the production of ROS in cells to maintain a relatively balanced concentration [22]. Nevertheless, the dysregulation of the activity of the antioxidant enzyme system may cause an imbalance between production and scavenging of ROS, leading to higher concentrations of ROS, resulting in damage of various cellular components, such as lipids, enzymes, proteins, cell membranes, and DNA [23–25]. Furthermore, ROS overproduction ultimately leads to oxidative stress and triggering of the apoptotic pathway [26,27]. The balance of production and scavenging of ROS is considered key to maintaining healthy biological systems. Accordingly, antioxidant defense mechanisms that consist of enzymatic antioxidants are used to neutralize the extra ROS, including CAT (catalase), SOD (superoxide dismutase), GR (glutathione reductase), and GPx (glutathione peroxidases) [27–29]. Furthermore, exogenous antioxidants are often used to inhibit the rate of ROS production, such as NAC (*N*-acetylcysteine), vitamin C, vitamin E, polyphenols, and carotenoids [30].

NAC is an exogenous reactive oxygen scavenger and a precursor of reduced glutathione (GSH), which can directly react with oxidative substances to reduce ROS content in the body. Studies have shown that NAC destroyed thiolated proteins to release free mercaptans, which have better antioxidant activity than NAC, and promoted the synthesis of glutathione and reduced proteins [31]. It has been reported that NAC treatment effectively reduced the increase in ROS level of GES-1 cells caused by *Helicobacter pylori*, as well as inhibited the ROS-mediated activation of the PI3K/Akt pathway and DNA damage [32]. Through starvation and paraquat stress tests, NAC treatment was observed to upregulate the transcription levels of key enzymes that resist ROS attack, such as phospholipid peroxide glutathione peroxidase and catalase [33]. Additionally, NAC clearance of $PM_{2.5}$-induced ROS can block cell apoptosis and restore the activity of the Nrf2 signaling pathway in human embryonic stem cells [34].

In the current research, we systematically elucidated the antioxidant effect of NAC in the regulation of cell survival toward PAT-induced intracellular ROS generation in HEK293 cells. More importantly, the potential molecular mechanisms, the relationship between the production and scavenging of ROS, and the effect of PAT on the apoptosis of HEK293 cells were investigated. Our research contributes to an understanding of the pathological role of ROS and provides theoretical support for NAC's inhibition of PAT-induced cytotoxicity.

2. Result

2.1. PAT Induced Cytotoxicity and Apoptosis in HEK293 Cells, Diminished by NAC

Reports have revealed that ROS overproduction plays a key part in PAT-induced cytotoxicity [12]. Next, we tested the effects of NAC, an ROS scavenger, on cytotoxicity in response to PAT. As exhibited in Figure 1A, compared with PAT treatment group, the cell viability significantly increased by 102.07%, 389.48%, and 412.21% when pretreated with 2, 4, and 10 mM NAC. Therefore, 4 mM NAC was chosen for subsequent experiments as an optimum dose. We next examined the effect of NAC on cytotoxicity using the LDH assay. The result demonstrated that PAT treatment exhibited a remarkable increase in

LDH leakage. In contrast, cotreatment with NAC significantly reduced the LDH leakage (Figure 1B).

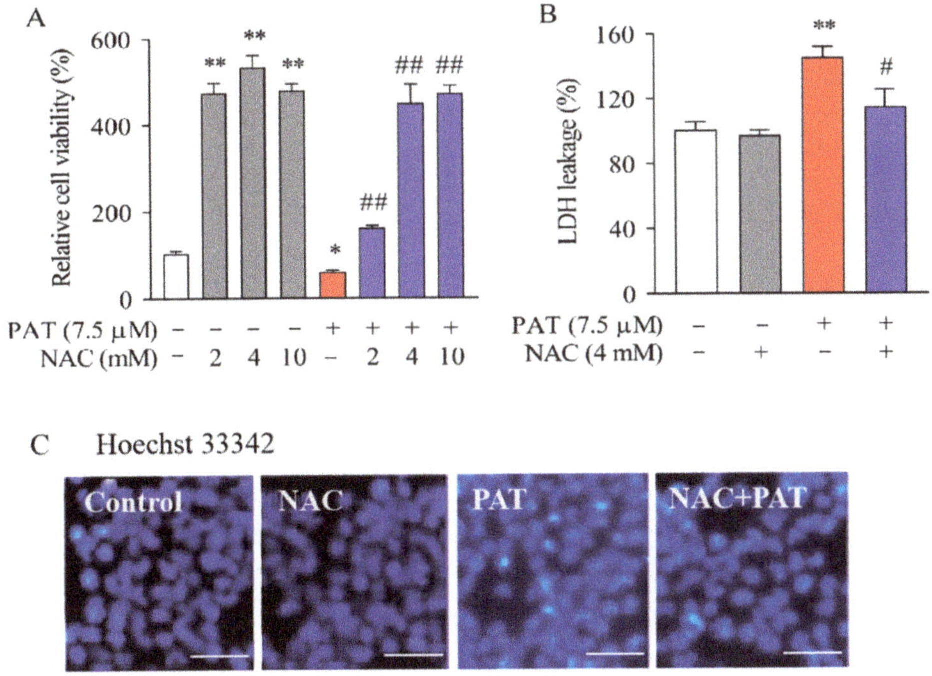

Figure 1. In HEK293 cells, NAC diminished the cytotoxicity and apoptosis caused by PAT. HEK293 cells were treated by 7.5 μM PAT with or without 4 mM NAC. (**A**) The cell viability of different groups was measured using an MTT assay (means ± SD, $n = 5$). (**B**) The cytotoxicity of different groups was measured using the LDH leakage assay (means ± SD, $n = 3$). (**C**) Hoechst 33342 was used to stain HEK293 cells. The images are representative of three independent experiments. $^{\#}$ $p < 0.05$ and $^{\#\#}$ $p < 0.01$ compared with the PAT treatment group; * $p < 0.05$ and ** $p < 0.01$ compared with the control treatment group.

Next, Hoechst 33342 (Beyotime Institute of Biotechnology, Beijing, China) staining was performed to measure cell apoptosis. When apoptosis occurs, the nucleus undergoes pyknosis. After staining with Hoechst 33342, the normal cells will appear normal blue under a fluorescence microscope, while the apoptotic cells will appear dense and heavily stained bright blue. As shown in Figure 1C, HEK293 cells exhibited a normal nucleus structure in the control and NAC alone treatment group. On the other hand, PAT treatment caused nuclear degradation and chromatin condensation in cells. Therefore, the Hoechst staining suggests apoptosis in PAT-treated cells due to nuclear morphological change. On the contrary, cotreatment with NAC obviously ameliorated the cytotoxicity induced by PAT.

To support this hypothesis, the apoptosis rate was further detected by annexin V and PI staining. Annexin V has a high affinity binding for PS, which can be utilized as a sensitive probe to explore PS exposed on the surface of cell membrane. The transfer of PS to the outer membrane may also occur during cell necrosis, but the cell membrane of necrotic cells is damaged, whereas the DNA of the cells can be stained by PI, while PI does not stain the early apoptotic cells [35]. On the scatter plot of bivariate flow cytometry, the upper right quadrant denotes annexin V+/PI+, which shows late-phase apoptotic cells, and the lower right quadrant refers to Annexin V+/PI−, which shows early-phase cells. These two quadrants were used to quantify the results in Figure 2B. As exhibited in Figure 2A,B, compared with the control group, PAT treatment significantly caused apoptosis. In contrast, cotreatment with NAC decreased the percentage of apoptosis cells. The results suggest that, in HEK293 cells, NAC effectively decreased the apoptosis caused by PAT.

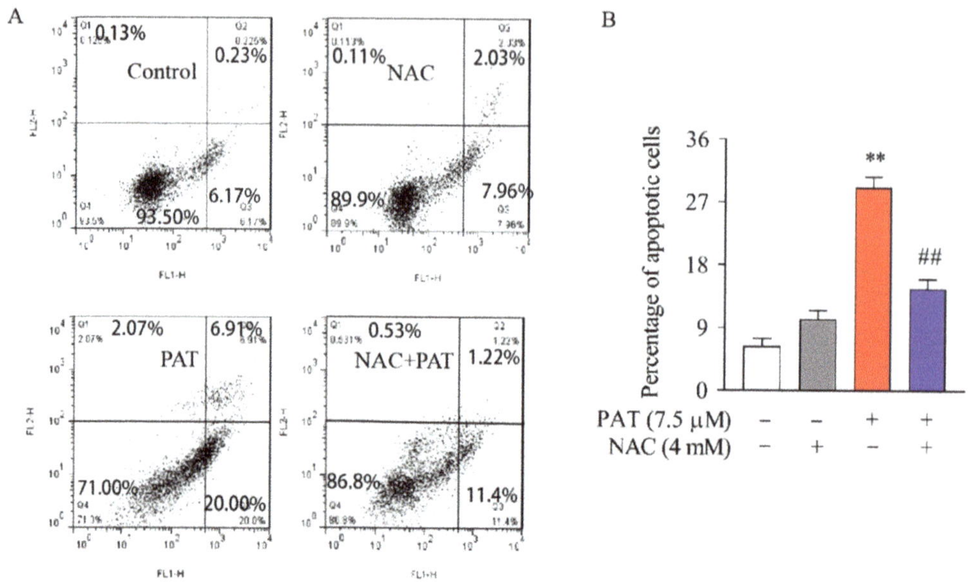

Figure 2. In HEK293 cells, NAC diminished cytotoxicity and apoptosis caused by PAT. HEK293 cells were treated by 7.5 μM PAT with or without 4 mM NAC. (**A**) Apoptosis was analyzed by flow cytometry with annexin V–FITC and PI dual staining. FL1 refers to annexin V–FITC and FL2 refers to PI. The lower right quadrant refers to annexin V–FITC-stained cells and the upper right quadrant refers to PI and annexin V–FITC-dual-stained cells. (**B**) Quantification of apoptosis rates (means ± SD, $n = 3$). ## $p < 0.01$ compared with the PAT treatment group; ** $p < 0.01$ compared with the control treatment group.

2.2. *NAC Reduced ROS Generation in Response to PAT Exposure*

To clarify the role of NAC in oxidative stress induced by PAT, the changes in intercellular ROS levels were measured by DCFH-DA probe. As exhibited in Figure 3A,B, the mitochondria-derived ROS was assessed by a MitoSOX Red Mitochondrial Superoxide Indicator. Results indicated that, compared to the control group, PAT significantly increased the intensity of red fluorescence. Furthermore, PAT treatment resulted in a remarkable increase in intracellular ROS compared with the control treatment. Nevertheless, the ROS level was dramatically decreased in the presence of NAC (Figure 3C). The changes were significantly attenuated by NAC. All these results showed that NAC obviously reduced the generation of total and mitochondria-derived ROS in response to PAT exposure.

2.3. *PAT Caused the Impairment of the Mitochondrial Respiratory Chain Complex Signaling Pathway*

It is known that the mitochondrial respiratory chain (MRC) is a crucial origin of ROS in eukaryotic cells [36]. To elucidate whether the ROS production induced by PAT is caused by the disorder of MRC, the mRNA and protein expressions of the MRC complexes were determined. As shown in Figure 4, treatment with PAT induced a concentration-dependent reduction in complex III (UQCRQ) and complex V (ATP6, ATP8) mRNA levels of HEK293 cells. However, PAT increased the mRNA expression of complex IV (COX17) in a concentration-dependent manner. In addition, the 2.5 μM PAT treatment increased the mRNA transcription of complex I (NDUFA4) and complex II (SDHA), while the 7.5 μM

PAT treatment decreased the mRNA transcription of complex II (SDHA). Similar results were observed in protein expressions of the MRC complexes (Figure 5A,B). These results suggested that PAT impaired the process of mitochondrial oxidative phosphorylation, leading to mitochondrial dysfunction and the generation of ROS.

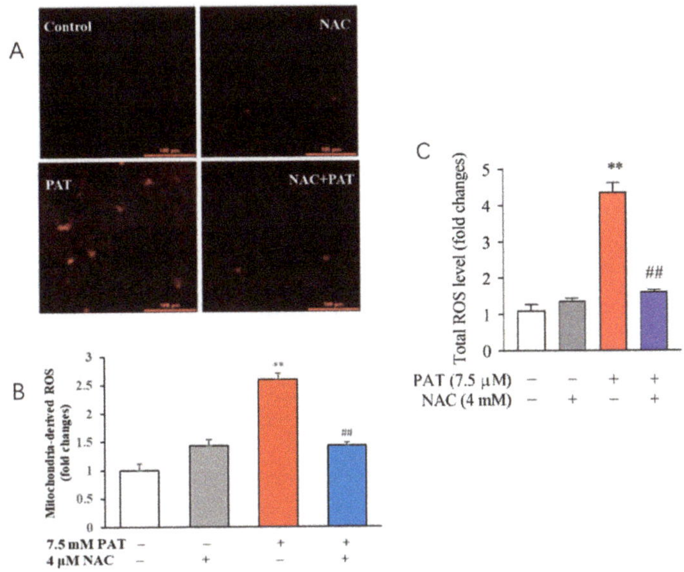

Figure 3. NAC reduced ROS reproduction in response to PAT exposure. HEK293 cells were treated by 7.5 μM PAT with or without 4 mM NAC. (**A**) Fluorescent image. The mitochondrial ROS level was measured with MitoSOX Red Mitochondrial Superoxide Indicator. The images are representative of three independent experiments. (**B**) Fold changes of mitochondrial ROS (means ± SD, $n = 3$). (**C**) The DCFH-DA probe was used to assess total ROS levels (means ± SD, $n = 3$). ## $p < 0.01$ compared with the PAT treatment group; ** $p < 0.01$ compared with the control treatment group.

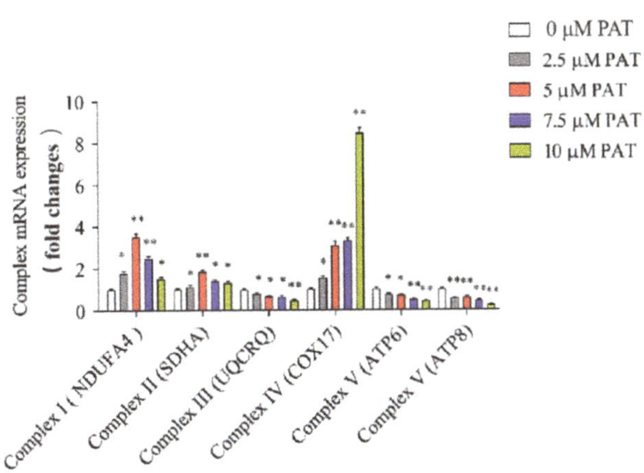

Figure 4. The gene levels of the mitochondrial respiratory chain complex. PAT caused the impairment of the mitochondrial respiratory chain complex signaling pathway. After 10 h treatment of HEK293 cells with different concentrations of PAT, the mRNA expression was detected by real-time PCR (means ± SD, $n = 3$). * $p < 0.05$ and ** $p < 0.01$ compared with the 0 μM PAT group.

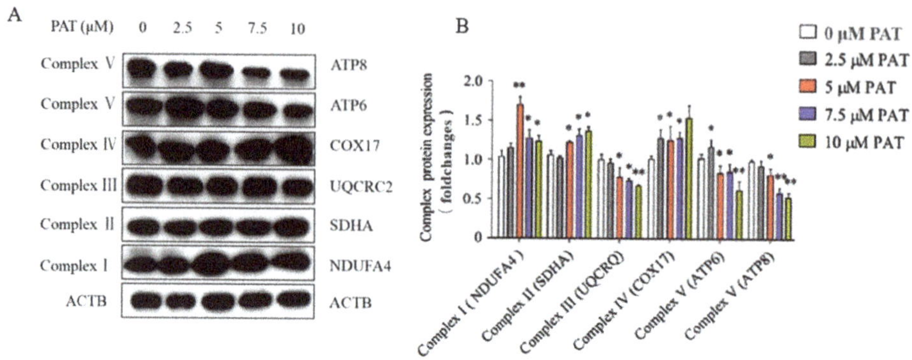

Figure 5. PAT caused the impairment of the mitochondrial respiratory chain complex signaling pathway. Different concentrations of PAT were used to treat HEK293 cells for 10 h. (**A**) Western blot detection of respiratory chain complex protein expression after PAT treatment at different concentrations. β-Actin served as an internal reference. The images are representative of three independent experiments. (**B**) Mitochondrial respiratory chain complex protein expressions (means ± SD, $n = 3$). * $p < 0.05$ and ** $p < 0.01$ compared with the 0 μM PAT group.

2.4. NAC Improved the Disorders of the Mitochondrial Respiratory Chain Complex Induced by PAT

Mitochondria are relevant to the increase in ROS, mainly through the mitochondrial electron transport chain. To verify the role of NAC in the mitochondrial dysfunction caused by PAT, the mRNA and protein expressions of the MRC complexes were measured. As shown in Figure 6, PAT treatment obviously increased the expression of mRNA in complex II (SDHA) and complex IV (COX17). In addition, NAC improved the complex II (SDHA) and complex IV (COX17) gene expression. In contrast, PAT decreased complex III (UQCRQ) and complex V (ATP6, and ATP8) gene expression, whereas it increased after cotreatment with NAC. Western blotting showed similar results to mRNA expression (Figure 7A,B). Altogether, these results showed that NAC attenuated the mitochondrial dysfunction caused by PAT via the MRC complex pathway.

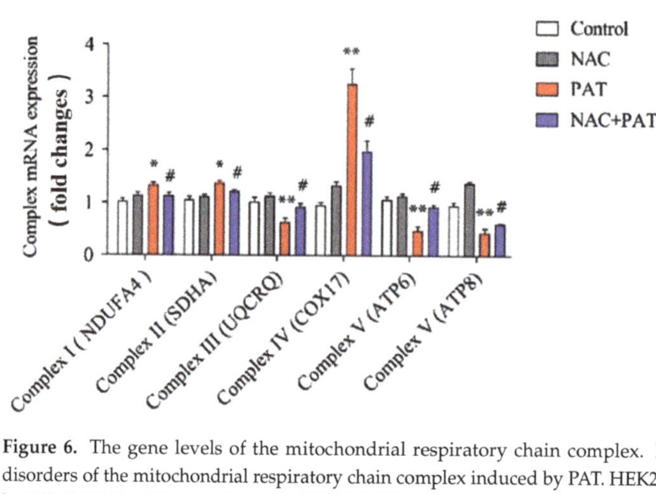

Figure 6. The gene levels of the mitochondrial respiratory chain complex. NAC improved the disorders of the mitochondrial respiratory chain complex induced by PAT. HEK293 cells were treated by 7.5 μM PAT with or without 4 mM NAC. The mRNA expression was determined by real-time PCR (means ± SD, $n = 3$). # $p < 0.05$ compared with the PAT treatment group; * $p < 0.05$ and ** $p < 0.01$ compared with the control treatment group.

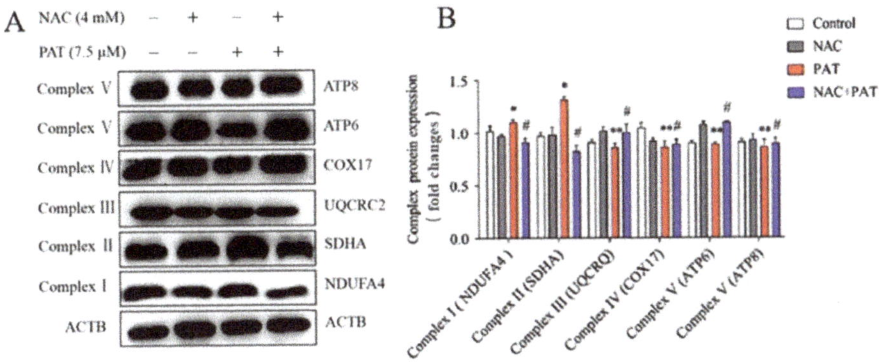

Figure 7. NAC improved the disorders of the mitochondrial respiratory chain complex induced by PAT. HEK293 cells were treated by 7.5 μM PAT with or without 4 mM NAC. (**A**) Western blot detection of respiratory chain complex protein expression after different treatments. β-Actin served as an internal reference. The pictures are representative of three independent experiments. (**B**) Mitochondrial respiratory chain complex protein expressions (means ± SD, $n = 3$). # $p < 0.05$ compared with the PAT treatment group; * $p < 0.05$ and ** $p < 0.01$ compared with the control treatment group.

2.5. NAC Protected against PAT-Induced GSH Depletion and Improved Antioxidant Enzyme Activities

Accumulating evidence has indicated that glutathione metabolism and other antioxidant enzymes are the most critical cellular defense mechanisms for protecting against oxidative stress damage [37]. To explore the positive effect of NAC, the levels of GSH and GSSG were examined. PAT treatment decreased GSH level and increased GSSG level significantly compared with the control group. In contrast, NAC pretreatment exhibited an opposite effect compared to the PAT group (Figure 8A,B). Consistently, the ratio of GSH to GSSG was significantly increased in the PAT plus NAC treatment group (Figure 8C). The activities of CAT, SOD, GPx, and GR were reduced notably after PAT treatment, whereas cotreatment with NAC dramatically increased their activities (Figure 8D–G). Taken together, these findings suggested that NAC can reduce the GSH depletion and restore the decline in antioxidant enzyme activities induced by PAT.

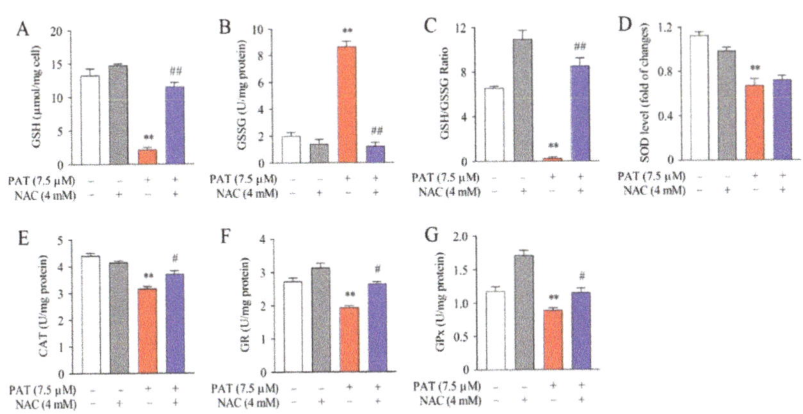

Figure 8. NAC protected against PAT-induced GSH depletion and improved antioxidant enzyme activities. HEK293 cells were treated by 7.5 μM PAT with or without 4 mM NAC. (**A**) Content of GSH; (**B**) content of GSSG; (**C**) ratio of GSH to GSSG; (**D**) activity of SOD; (**E**) activity of CAT; (**F**) activity of GR; (**G**) activity of GPx. All the above were measured using different assay kits (means ± SD, $n = 3$). # $p < 0.05$ and ## $p < 0.01$ compared with the PAT treatment group; ** $p < 0.01$ compared with the control treatment group.

2.6. NAC Inhibited PAT-Induced Mitochondrial Dysfunction and Caspase-Dependent Apoptotic Pathway

The results showed that MMP was decreased in the cells treated by PAT; the green fluorescence was enhanced while the red fluorescence was weakened. However, the value of the MMP was attenuated by cotreatment with NAC (Figure 9A,B). We next examined the ATP level using an ATP assay kit. As exhibited in Figure 9C, PAT prominently reduced ATP production when compared with control cells. In contrast, it exhibited a notable increase in ATP levels upon pretreatment with NAC. Taken together, these results revealed that NAC suppressed PAT-induced mitochondrial dysfunction.

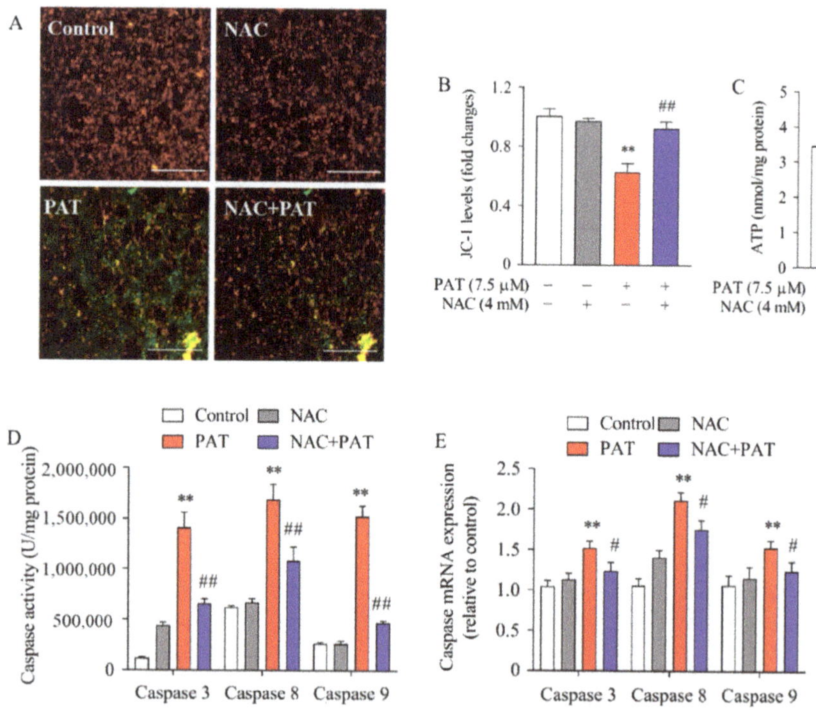

Figure 9. NAC inhibited PAT-Induced mitochondrial dysfunction and apoptosis dependent on caspase pathway. HEK293 cells were treated by 7.5 μM PAT with or without 4 mM NAC. (**A**,**B**) JC-1 staining was used to analyze MMP. Fluorescent images are representative of three independent experiments. JC-1 levels came from the ratio of red/green fluorescence (means ± SD, $n = 3$). (**C**) ATP levels of different treatments were detected using an ATP assay kit (means ± SD, $n = 3$). (**D**) The activities of the caspase 3, caspase 8, and caspase 9 were measured as described in Section 5 (means ± SD, $n = 3$). (**E**) Real-time PCR analyses for the expression of caspase 3, caspase 8, and caspase 9 (means ± SD, $n = 3$). $^{\#}$ $p < 0.05$ and $^{\#\#}$ $p < 0.01$ compared with the PAT treatment group; ** $p < 0.01$ compared with the control treatment group.

Caspases represent a family of proteolytic enzymes that function as initiators and executors of apoptosis [38]. Our previous research showed that apoptosis induced by PAT was related to caspase cascade activation [39]. Therefore, we assessed the activities and mRNA expressions of caspase 3, caspase 8, and caspase 9. As exhibited in Figure 9D, the activities of caspases 3, 8, and 9 obviously increased after PAT treatment, whereas pretreatment with NAC caused a significant reduction in caspase 3, 8, and 9 activities. Consistent with the enzymatic activities, the mRNA levels of caspase 3, 8, and 9 were greatly upregulated after PAT treatment, whereas they were significantly attenuated by

NAC (Figure 9E). Collectively, these results clearly indicate that NAC suppressed PAT-induced apoptosis through the caspase-dependent pathway.

3. Discussion

Studies have evidenced that PAT exerts cytotoxicity through accelerating the production of intracellular ROS, leading to apoptosis [11,40]. Mitochondrial oxidative phosphorylation is the main source of ROS, which is made up of a sequence of respiratory complexes (complexes I–V) [41,42], and alterations in oxidative phosphorylation (OXPHOS) result in the generation of ROS, further inducing oxidative stress and mitochondrial dysfunction [13,43–45].

In the present research, we observed a significant change in the activities of mitochondrial respiratory complexes in response to PAT. In detail, the levels of complex III (UQCRQ) and complex V (ATP6, ATP8) obviously decreased, whereas that of complex IV (COX17) increased after PAT treatment; however, the levels of complex I (NDUFA4) and complex II (SDHA) first increased and then decreased with an increase in the concentration of PAT. Moreover, total and mitochondrial ROS increased greatly.

Our previous studies have indicated that PAT can induce apoptosis [12,39]. The current study focused on the protective role of NAC, a water-soluble antioxidant agent, on PAT-induced apoptosis in HEK293 cells. Studies have shown that NAC can scavenge free radicals, increase the level of cellular glutathione, and decrease depolarization of the mitochondrial membrane [46,47]. In this study, the production of total and mitochondrial ROS was notably attenuated by NAC treatment. Briefly, NAC treatment elevated the ATP contents and MMP values compared with the PAT treatment group. Moreover, our studies found that inhibition of antioxidant NAC improved mitochondrial oxidative phosphorylation through the regulation of MRC complexes.

It is well known that caspases are involved in apoptosis regulation. In the current research, we discovered a significant increase in the enzyme activities of caspases 3, 8, and 9 in response to PAT exposure, whereas NAC effectively inhibited their activities. Consistent with this result, NAC led to a significant decrease in the mRNA levels of caspases 3, 8, and 9. These results indicated that NAC reduce PAT-induced apoptosis by inhibiting the activation of caspases 3, 8, and 9.

To protect cells from oxidative damage, cellular defense mechanisms will produce numerous antioxidant enzymes (SOD, CAT, GR, GPx) and antioxidants such as GSH to neutralize superoxide radicals. It is well known that GSH can be oxidized to GSSG. Consistent with our previous studies, we discovered that intracellular ROS overproduction induced by PAT is accompanied by GSH depletion and increased GSSG content [39]. However, in the current research, we noticed marked increases in the ratio of GSH to GSSG involved in groups administered NAC and PAT. To further verify the protective effects of NAC, we measured the activities of SOD, CAT, GR, and GPx. Research has shown that the cellular antioxidant enzyme SOD plays a central role in scavenging the superoxide ion by speeding up its dismantlement [37], whereas CAT is a familiar enzyme which can convert one hydrogen peroxide molecule into two independent water molecules through catalyzing peroxide reactions [48]. GR plays a crucial part in the neutralization of hydroperoxides; moreover, GPx can scavenge hydrogen peroxide and other peroxides using GSH as the reducing agent [49–51]. The results obtained in the present research also proved that pretreatment with NAC obviously reversed the decrease in activities of SOD, CAT, GR, and GPx induced by PAT. Taken together, our results clearly suggest that NAC can obviously attenuate the PAT-induced oxidative damage through decreasing the activity of antioxidant enzymes.

At present, most studies on the toxicity of PAT have been conducted in vitro. Existing in vivo experiments have shown that feeding 10 mg/kg PAT for 4 days in BALB/c mice significantly increased the levels of serum urea and LDH, and some pathological changes including renal tubule swelling, vacuolar degeneration, and protein casting occurred in the kidneys [3]. BALB/C mice were fed 2 mg/kg PAT for 7 consecutive days and then injected intraperitoneal with the same concentration for 3 days; the production of ROS in liver tissue measured by DCFH-DA increased by 288%, the expression of oxidative markers such as SOD and CAT was downregulated, and the expression of apoptotic enzymes p53 and caspase 3 was increased. The results indicated that PAT promoted oxidative stress and apoptosis in the liver [52]. The toxicity of PAT demonstrated by the results of this study is consistent with the above in vivo experiments, which lays a foundation for further study of the effect of NAC on the ROS-mediated oxidative damage pathway induced by PAT in vivo.

4. Conclusions

In summary, our study demonstrated that PAT can promote mitochondrial dysfunction and oxidative stress in HEK293 cells through increasing intracellular ROS production and regulating mitochondrial function. Furthermore, treatment with the antioxidant NAC significantly suppressed oxidative damage and apoptosis in response to PAT exposure. Together, these findings can help to understand fundamental molecular mechanisms underlying ROS production and apoptosis for the treatment of PAT, as well as provide a novel treatment strategy for PAT-induced cytotoxicity.

5. Materials and Methods

5.1. Chemicals and Reagents

N-Acetyl-1-cysteine (NAC, purity $\geq$ 98%) and patulin (PAT, purity $\geq$ 99%) were supplied by Sigma-Aldrich (St. Louis, MO, USA). The LDH-assay kit, ROS assay kit, total SOD assay kit, annexin V–FITC apoptosis assay kit, Hoechst 33342 dyes, GSH and GSSG assay kit, GR assay kit, total GPx assay kit, ATP assay kit, mitochondrial membrane potential (MMP) assay kit, the assay kits of caspase 3, 8, and 9, and BCA protein-assay kit were obtained from Beyotime Institute of Biotechnology (Beijing, China). MitoSOX Red Mitochondrial Superoxide Indicator was purchased from Invitrogen Corporation (St. Louis, MO, USA). The CAT test kit was obtained from Jiancheng Bioengineering Institute (Nanjing, Jiangsu, China). The Ultrapure RNA kit, Super RT cDNA kit, and UltraSYBR mixture were purchased from CWBIO (Beijing, China).

5.2. Cell Cultivation and Treatment

HEK293 cells were purchased from Zhongqiao Xinzhou Biotechnology Company (Shanghai, China), maintained in Dulbecco's modified Eagle medium (DMEM) supplemented with 10% FBS and 1% penicillin–streptomycin. At the temperature of 37 °C, the cells were cultivated in a humidified incubator with 5% CO_2 and 95% air. When the HEK293 cells were grown to approximately 80% confluence, the cells were treated with 7.5 μM PAT for 10 h, or the cells were pretreated with NAC at different concentrations (0, 2, 4, and 10 mM) for 3 h.

5.3. Cell Viability Assays

MTT assay and LDH leakage were used to determine the cell viability according to the manufacturer's instructions. In short, 1×10^5 HEK293 cells were plated in a 96-well culture plate. Then, 24 h later, 7.5 μM PAT was added to the cells, or they were pretreated with different concentrations of NAC (0, 2, 4, and 10 mM) for 3 h, before treating with PAT for 10 h. Thereafter, 10 μL of MTT reagent (5 mg/mL in sterile PBS) was added to every well and incubated at 37 °C for 4 h. Next, the medium was discarded, and the sample was supplemented 150 μL of DMSO to dissolve formazan crystals. A microplate reader (Bio-Rad 680) was used to measure the absorbance at a wavelength of 570 nm. All experiments were repeated five times.

The LDH assay kit was used to assess LDH leakage according to the manufacturer's instructions. PAT (7.5 μM) was added to cells, or 4 mM NAC was added to pretreat the cells for 3 h, before being treated by PAT for 10 h. The absorbance was measured using a microplate reader (Bio-Rad 680) at 490 and 630 nm wavelengths.

5.4. Hoechst 33342 Staining

Hoechst 33342 staining was used to examine cell apoptosis. After treating with 7.5 μM PAT or pretreating with 4 mM NAC for 3 h, the cells were treated with PAT for 10 h. The cells were washed twice with PBS (pH 7.4) and then incubated with Hoechst 33342 dye (10 μg/mL) in the dark for 20 min (at the condition of 37 °C). The cells were observed under an inverted fluorescence microscope (Lecia DMI8, Germany).

5.5. Apoptosis Assessment

Flow cytometry with an annexin V–FITC/PI apoptosis detection kit was used to detect the cell apoptosis ratio. Briefly, the HEK293 cells were seeded in a six-well plate, treated as described above. Then, the cells were trypsinized, collected, and washed twice with ice-cold PBS, before centrifuging at 1000× *g* for 5 min. The collected cells were resuspended with 195 μL of annexin V–FITC binding buffer, and stained with 5 μL of annexin V–FITC for 10 min and 10 μL of propidium iodide (PI) for 5 min in the dark. Afterward, the cells were immediately analyzed by flow cytometry (FACS Calibur, Franklin Lakes, NJ, USA).

5.6. Measurement of Mitochondrial Membrane Potential

The JC-1 fluorescent probe was used to assess the mitochondrial membrane potential. In short, after treatment as described above, the cells were cultured with the JC-1 dye for 20 min at a temperature of 37 °C. After that, the stained cells were washed twice with JC-1 dyeing buffer (1×). Then, the samples were observed under inverted fluorescence microscopy. The red/green fluorescence values were determined using a multifunctional microplate analyzer (Tecan, Infinite M200 Pro, Männedorf, Switzerland).

5.7. Measurement of ATP Level

An ATP assay kit was used to determine the adenosine triphosphate content according to the manufacturer's protocols. After treatment as above, the cells were broken down and centrifuged at 12,000× *g* for 5 min at a temperature of 4 °C to collect supernatant. Then, 10 μL of supernatant was mixed with 100 μL of ATP detection solution in black 96-well plates. After incubation for 5 min, the change in ATP levels was detected by chemiluminescence using a multifunctional microplate reader (Tecan, Infinite M200 Pro, Männedorf, Switzerland).

5.8. Measurement of Total ROS and Mitochondrial ROS

Total ROS in HEK293 cells was evaluated using a 2,7-dichloroflfluorescein diacetate (DCFH-DA) probe. After treatment as above, the cells were cultured using a 10 μM DCFH-DA probe for 20 min at 37 °C. The fluorescence was observed under a fluorescent microscope. The fluorescence intensity was measured using a multifunctional microplate analyzer at 488 nm excitation wavelength and 520 nm emission wavelength.

To measure the mitochondrial ROS, the MitoSOX Red Mitochondrial Superoxide Indicator was used according to the manufacturer's instructions. After treatment as described above, the working probe fluid was added to react with cells for 10 min at a temperature of 37 °C. Then, a laser confocal microscope (Nikon, A1) was used to obtain the fluorescence images.

5.9. Analysis of GSH and GSSG Contents

GSH and GSSG assay kits were used to measure GSH and GSSG levels. Briefly, after treatment, the cells were washed with ice-cold PBS, before being centrifuged at 600× *g* for 5 min. Next, the cell volume was evaluated. Then, a protein-removal regent at 10 times the cell volume was added, and the cells underwent two freeze–thaw cycles in liquid nitrogen and a 37 °C water bath. The cells were centrifuged at 10,000× *g* for 10 min at 4 °C to collect supernatant. Afterward, the GSH and GSSG content was measured according to the manufacturer's instructions. The absorbance at 412 nm was assayed using a multifunctional microplate reader.

5.10. Activities of SOD, CAT, GR, and GPx

Briefly, cells were treated as above, and the supernatants were collected after lysing and centrifuging at 12,000× *g* at 4 °C for 10 min. The protein concentration was detected using a BCA protein assay kit.

Activity of SOD: A total superoxide dismutase assay kit with WST-8 was used to measure SOD activity. The cell supernatants were incubated with WST working solution and enzyme working solution at room temperature for 20 min. A multifunctional microplate reader was used to measure absorbance at 450 nm.

Activity of CAT: A catalase assay kit was used to detect the CAT content according to our previous studies [53].

Activity of GR: GR activity was measured using a glutathione reductase assay kit with NADPH as the reducing agent. The amount of yellow-TNB was used to reflect GR activity, produced by the reaction between GSH and DTNB. A multifunctional microplate reader was used to measure absorbance at 412 nm.

Activity of GPx: GPx activity was determined using a total glutathione peroxidase assay kit. GPx activity was measured on the basis of the decrease in NADPH in the presence of GPx. A multifunctional microplate reader was used to measure absorbance at 340 nm.

5.11. Caspase Activity Assay

The corresponding caspase activity assay kits were used to assay the activities of caspases 3, 8, and 9 according to the manufacturer's instructions. In short, the cells were treated as described above. The cells were collected and lysed with lysis buffer, before being centrifuged at 16,000× *g* for 15 min at a temperature of 4 °C. Afterward, the supernatants were incubated with 5 μL of Ac-DEVD-pNA, Ac-IETD-pNA, or Ac-LEHD-pNA in the dark at 37 °C for 2 h. The absorbance at 405 nm was determined using a microplate reader (Bio-Rad 680).

5.12. Quantitative Real-Time Reverse Transcription PCR (qRT-PCR)

After treatment as mentioned above, TRIzon Reagent was used to extract total mRNA from HEK293 cells. The RNA was reverse-transcribed into cDNA using the Super RT cDNA kit according to the manufacturer's protocol. In the qRT-PCR analysis, cDNA amplification was performed using a two-step UltraSYBR mixture and was detected by the IQ5 Multicolor Real-Time PCR Detection System (Bio-Rad). The primer sequences are provided in Appendix A.

5.13. Western Blot Analysis

The Western blot assay was processed as described previously (Wang et al., 2018). Briefly, the cells were broken down using a lysis buffer supplemented with 1 mM PMSF and then homogenized for 10 min (11,000× *g*, 4 °C). The supernatants were collected and then denatured with 4× loading buffer at 95 °C for 10 min. The BCA protein assay kit was used to determine the protein concentration. Subsequently, the proteins were separated by 10–15% SDS-PAGE, and then transferred onto a PVDF membrane, which was blocked with 5% nonfat milk dissolved in TBST at room temperature for 2 h. Next, they were incubated with primary antibodies (1:1000), including anti-NDUFA4, anti-SDHA, anti-UQCRQ, anti-COX17, anti-ATP6V1B2, anti-ATP8B2, and anti-β-actin. After incubation overnight at 4 °C, the blots were washed three times for 20 min before incubating with the secondary antibody at room temperature for 2 h. The blots were detected using a hypersensitive ECL chemiluminescence reagent (Beyotime) and exposed to a Chemiluminescent imaging system (Bio-Rad). Image J software was used to quantitate the protein intensity.

5.14. Statistical Analysis

All data are presented as means ± standard deviations (SD) from at least three independent experiments. Significant differences were analyzed with one-way factorial analysis of variance (ANOVA) and Tukey's test (SPSS 19.0). A value of $p < 0.05$ represented a statistically significant difference.

Author Contributions: Conceptualization, J.L. and X.P.; methodology, J.L., Q.L., and J.F.; formal analysis, J.L., Q.L., and R.J.; writing—original draft preparation, J.L.; writing—review and editing, J.L.; software, J.H. and Z.L.; investigation, J.L., Q.L., and J.H.; validation, Q.L., J.F., F.M., and R.J.; visualization, J.L. and T.G.; supervision, X.P.; resources, X.P.; project administration, X.P.; funding acquisition, X.P. All authors have read and agreed to the published version of the manuscript.

Funding: This research was funded by grants from the National Natural Science Foundation of China (NSFC, 31571928) and the Scientific and Technological Project of Shaanxi Province (2020NY-106).

Conflicts of Interest: The authors declare no conflict of interest.

Appendix A

Table A1. Sequences and parameters of the primers used in real-time PCR.

Gene Name	Forward Primer	Reverse Primer	Amplicon Size (bp)
ACTB	TCGTGCGTGACATTAAGGAG	AGGAAGGAAGGCTGGAAGAG	178
NDUFA4	AGCTTGATCCCCCTCTTTGT	TCATTGGGACCCAGTTTGTT	140
SDHA	CCTACCAGGTCACACACTGT	AGTCCGATCAGCCACACAG	162
UQCRC2	TCAAGAACTGGTCCCGAC	TGTAAGGTACCCAGTCCAGG	102
COX17	TAGATTTGGCTGTCTCCGCT	CTCTCATGCATTCCTTGTGGG	213
ATP6	GGACTCCTGCCTCACTCATT	ATAAGGGGTGTAGGTGTGCC	158
ATP8	ATGGCCCACCATAATTACCC	GCAATGAATGAAGCGAACAG	170
Caspase 3	TCCTGAGATGGGTTTATGT	ATGTTTCCCTGAGGTTTGC	110
Caspase 8	CTCCAAATGCAAACTGGATG	TGTTGATTTGGGCACAGACT	122
Caspase 9	CCGCATACTCCATCCTCAGT	GCTTCTTCATCCCAAAGTCG	123

References

1. Braun, H.P. The oxidative phosphorylation system of the mitochondria in plants. *Mitochondrion* **2020**, *53*, 66–75. [CrossRef] [PubMed]
2. Liu, B.H.; Wu, T.S.; Yu, F.Y.; Su, C.C. Induction of oxidative stress response by the mycotoxin patulin in mammalian cells. *Toxicol. Sci.* **2007**, *95*, 340–347. [CrossRef] [PubMed]
3. Lu, X.; Zhang, E.; Yin, S.; Fan, L.; Hu, H. Methylseleninic acid prevents patulin-induced hepatotoxicity and nephrotoxicity via the inhibition of oxidative stress and inactivation of p53 and MAPKs. *J. Agric. Food Chem.* **2017**, *65*, 5299–5305. [CrossRef]
4. Boussabbeh, M.; Ben Salem, I.; Prola, A.; Guilbert, A.; Bacha, H.; Abid-Essefi, S.; Lemaire, C. Patulin induces apoptosis through ROS-mediated endoplasmic reticulum stress pathway. *Toxicol. Sci.* **2015**, *144*, 328–337. [CrossRef]
5. Jin, H.; Yin, S.; Song, X.; Zhang, E.; Fan, L.; Hu, H. p53 activation contributes to patulin-induced nephrotoxicity via modulation of reactive oxygen species generation. *Sci. Rep.* **2016**, *6*, 24455. [CrossRef] [PubMed]
6. Liu, B.H.; Wu, T.S.; Yu, F.Y.; Wang, C.H. Mycotoxin patulin activates the p38 kinase and JNK signaling pathways in human embryonic kidney cells. *Toxicol. Sci.* **2006**, *89*, 423–430. [CrossRef]
7. Guerra-Moreno, A.; Hanna, J. Induction of proteotoxic stress by the mycotoxin patulin. *Toxicol. Lett.* **2017**, *276*, 85–91. [CrossRef]
8. Pillay, Y.; Phulukdaree, A.; Nagiah, S.; Chuturgoon, A.A. Patulin triggers NRF2-mediated survival mechanisms in kidney cells. *Toxicon* **2015**, *99*, 1–5. [CrossRef]
9. Puel, O.; Galtier, P.; Oswald, I.P. Biosynthesis and toxicological effects of patulin. *Toxins* **2010**, *2*, 613–631. [CrossRef]
10. Kunio, D.; Koji, U. Mechanisms of mycotoxin-induced dermal toxicity and tumorigenesis through oxidative stress-related pathways. *J. Toxicol. Pathol.* **2014**, *27*, 1–10.
11. Boussabbeh, M.; Prola, A.; Ben Salem, I.; Guilbert, A.; Bacha, H.; Lemaire, C.; Abis-Essefi, S. Crocin and quercetin prevent PAT-induced apoptosis in mammalian cells: Involvement of ROS-mediated ER stress pathway. *Environ. Toxicol.* **2016**, *31*, 1851–1858. [CrossRef] [PubMed]
12. Zhang, B.; Peng, X.; Li, G.; Xu, Y.; Xia, X.; Wang, Q. Oxidative stress is involved in Patulin induced apoptosis in HEK293 cells. *Toxicon* **2015**, *94*, 1–7. [CrossRef]
13. Grancara, S.; Zonta, F.; Ohkubo, S.; Brunati, A.M.; Agostinelli, E.; Toninello, A. Pathophysiological implications of mitochondrial oxidative stress mediated by mitochondriotropic agents and polyamines: The role of tyrosine phosphorylation. *Amino Acids* **2015**, *47*, 869–883. [CrossRef] [PubMed]
14. Mancuso, M.; Filosto, M.; Orsucci, D.; Siciliano, G. Mitochondrial DNA sequence variation and neurodegeneration. *Hum. Genom.* **2008**, *3*, 71–78. [CrossRef] [PubMed]

15. Pan, Y.; Cao, M.; Liu, J.; Yang, Q.; Miao, X.; Go, V.L.W.; Lee, P.W.N.; Xiao, G.G. Metabolic regulation in mitochondria and drug resistance. *Mitochondrial DNA Dis.* **2017**, *1038*, 149–171.
16. Castellanos, E.; Lanning, N.J. Phosphorylation of OXPHOS machinery subunits: Functional implications in cell biology and disease. *Yale J. Biol. Med.* **2019**, *92*, 523–531. [PubMed]
17. Xu, Q.; Vu, H.; Liu, L.; Wang, T.C.; Schaefer, W.H. Metabolic profiles show specific mitochondrial toxicities in vitro in myotube cells. *J. Biomol. NMR* **2011**, *49*, 207–219. [CrossRef]
18. Yang, Y.; Karakhanova, S.; Hartwig, W.; D'Haese, J.G.; Philippov, P.P.; Werner, J.; Bazhin, A.V. Mitochondria and mitochondrial ROS in cancer: Novel targets for anticancer therapy. *J. Cell. Physiol.* **2016**, *231*, 2570–2581. [CrossRef]
19. Cecarini, V.; Gee, J.; Fioretti, E.; Amici, M.; Angeletti, M.; Eleuteri, A.M.; Keller, J.N. Protein oxidation and cellular homeostasis: Emphasis on metabolism. *Biochim. Biophys. Acta (BBA)-Mol. Cell Res.* **2007**, *1773*, 93–104. [CrossRef]
20. Kietzmann, T. Cellular redox compartments. *Antioxid. Redox Signal.* **2019**, *30*, 1–4. [CrossRef]
21. Qi, J.; Wang, J.; Gong, Z.; Zhou, J.M. Apoplastic ROS signaling in plant immunity. *Curr. Opin. Plant Biol.* **2017**, *38*, 92–100. [CrossRef] [PubMed]
22. Kala, M.; Shaikh, M.V.; Nivsarkar, M. Equilibrium between anti-oxidants and reactive oxygen species: A requisite for oocyte development and maturation. *Reprod. Med. Biol.* **2017**, *16*, 28–35. [CrossRef] [PubMed]
23. Aitken, R.J.; Gibb, Z.; Baker, M.A.; Drevet, J.; Gharagozloo, P. Causes and consequences of oxidative stress in spermatozoa. *Reprod. Fertil. Dev.* **2016**, *28*, 1–10. [CrossRef] [PubMed]
24. Chen, Y.; Xiong, S.; Zhao, F.; Lu, X.; Wu, B.; Yang, B. Effect of magnesium on reducing the UV-induced oxidative damage in marrow mesenchymal stem cells. *J. Biomed. Mater. Res. Part A* **2019**, *107*, 1253–1263. [CrossRef] [PubMed]
25. Sabeti, P.; Pourmasumi, S.; Rahiminia, T.; Akyash, F.; Talebi, A.R. Etiologies of sperm oxidative stress. *Int. J. Reprod. BioMed.* **2016**, *14*, 231–240. [CrossRef] [PubMed]
26. Kajarabille, N.; Latunde-Dada, G.O. Programmed cell-death by ferroptosis: Antioxidants as mitigators. *Int. J. Mol. Sci.* **2019**, *20*, 4968. [CrossRef]
27. Cui, B.; Zhang, X.; Han, G.; Li, K. Antioxidant defense response and growth reaction of amorpha fruticosa seedlings in petroleum-contaminated soil. *Water Air Soil Pollut.* **2016**, *227*, 121. [CrossRef]
28. Ferrareze, J.P.; Fugate, K.K.; Bolton, M.D.; Deckard, E.L.; Campbell, L.G.; Finger, F.L. Jasmonic acid does not increase oxidative defense mechanisms or common defense-related enzymes in postharvest sugarbeet roots. *Postharvest Biol. Technol.* **2013**, *77*, 11–18. [CrossRef]
29. Krishnan, N.; Kodrik, D.; Turanli, F.; Sehnal, F. Stage-specific distribution of oxidative radicals and antioxidant enzymes in the midgut of Leptinotarsa decemlineata. *J. Insect Physiol.* **2007**, *53*, 67–74. [CrossRef]
30. Akhtar, M.J.; Ahamed, M.; Alhadlaq, H.A.; Alshamsan, A. Mechanism of ROS scavenging and antioxidant signalling by redox metallic and fullerene nanomaterials: Potential implications in ROS associated degenerative disorders. *Biochim. Biophys. Acta (BBA)-Gen. Subj.* **2017**, *1861*, 802–813. [CrossRef]
31. Giancarlo, A.; Alessandra, A.; Giovanna, B.; Giulio, V.; Marina, C.; Luisa, B.; Francesco, S. N-Acetylcysteine as an antioxidant and disulphide breaking agent: The reasons why. *Free Radic. Res.* **2018**, *52*, 751–762.
32. Xie, C.; Yi, J.; Lu, J.; Nie, M.; Huang, M.; Rong, J.; Zhu, Z.; Chen, J.; Zhou, X.; Li, B.; et al. N-Acetylcysteine reduces ROS-mediated oxidative DNA damage and PI3K/Akt pathway activation induced by helicobacter pylori Infection. *Oxidative Med. Cell. Longev.* **2018**, *2018*, 1874985. [CrossRef]
33. Niraula, P.; Kim, M.S. N-Acetylcysteine extends lifespan of Drosophila via modulating ROS scavenger gene expression. *Biogerontology* **2019**, *20*, 533–543. [CrossRef]
34. Jin, L.; Ni, J.; Tao, Y.; Weng, X.; Zhu, Y.; Yan, J.; Hu, B. N-acetylcysteine attenuates PM 2.5-induced apoptosis by ROS-mediated Nrf2 pathway in human embryonic stem cells. *Sci. Total Environ.* **2019**, *666*, 713–720. [CrossRef] [PubMed]
35. Lisa, C.C.; Brooke, J.M.; Adrian, P.S.; Nigel, J.W. Quantitation of apoptosis and necrosis by Annexin V binding, Propidium Iodide uptake, and flow cytometry. *Cold Spring Harb. Protoc.* **2016**, *11*, pdb-prot087288.
36. Droese, S.; Brandt, U. Molecular mechanisms of superoxide production by the mitochondrial respiratory chain. *Mitochondrial Oxidative Phosphorylation* **2012**, *748*, 145–169.
37. Carrillo-Vico, A.; Murillo-Cabezas, F.; Egea-Guerrero, J.J.; Rodriguez-Rodriguez, A. Oxidative stress in traumatic brain injury. *Curr. Med. Chem.* **2014**, *21*, 1201–1211.
38. Kopeina, G.S.; Prokhorova, E.A.; Lavrik, I.N.; Zhivotovsky, B. Alterations in the nucleocytoplasmic transport in apoptosis: Caspases lead the way. *Cell Prolif.* **2018**, *51*, e12467–e12469. [CrossRef]
39. Wang, X.; Jin, C.; Zhong, Y.; Li, X.; Han, J.; Xue, W.; Wu, P.; Xia, X.; Peng, X. Glutathione reduction of patulin-evoked cytotoxicity in HEK293 cells by the prevention of oxidative damage and the mitochondrial apoptotic pathway. *J. Agric. Food Chem.* **2018**, *66*, 7775–7785. [CrossRef]
40. Ramalingam, S.; Bahuguna, A.; Kim, M. The effects of mycotoxin patulin on cells and cellular components. *Trends Food Sci. Technol.* **2019**, *83*, 99–113. [CrossRef]
41. Mai, N.; Chrzanowska-Lightowlers, Z.M.A.; Lightowlers, R.N. The process of mammalian mitochondrial protein synthesis. *Cell Tissue Res.* **2017**, *367*, 5–20. [CrossRef]
42. Signes, A.; Fernandez-Vizarra, E. Assembly of mammalian oxidative phosphorylation complexes I-V and supercomplexes. *Essays Biochem.* **2018**, *62*, 255–270. [PubMed]

43. DiMauro, S.; Schon, E.A. Mechanisms of disease: Mitochondrial respiratory-chain diseases. *N. Engl. J. Med.* **2003**, *348*, 2656–2668. [CrossRef]
44. Du, S.N.N.; Khajali, F.; Dawson, N.J.; Scott, G.R. Hybridization increases mitochondrial production of reactive oxygen species in sunfish. *Evolution* **2017**, *71*, 1643–1652. [CrossRef]
45. Wojtala, A.; Karkucinska-Wieckowska, A.; Sardao, V.A.; Szczepanowska, J.; Kowalski, P.; Pronicki, M.; Duszynski, J.; Wieckowski, M.R. Modulation of mitochondrial dysfunction-related oxidative stress in fibroblasts of patients with Leigh syndrome by inhibition of prooxidative p66Shc pathway. *Mitochondrion* **2017**, *37*, 62–79. [CrossRef]
46. Mokhtari, V.; Afsharian, P.; Shahhoseini, M.; Kalantar, S.M.; Moini, A. A review on various uses of N-acetyl cysteine. *Cell J.* **2017**, *19*, 11–17. [PubMed]
47. Yamada, M.; Tsukimura, N.; Ikeda, T.; Sugita, Y.; Att, W.; Kojima, N.; Kubo, K.; Ueno, T.; Sakurai, K.; Ogawa, T. N-acetyl cysteine as an osteogenesis-enhancing molecule for bone regeneration. *Biomaterials* **2013**, *34*, 6147–6156. [CrossRef] [PubMed]
48. Samarghandian, S.; Azimi-Nezhad, M.; Farkhondeh, T.; Samini, F. Anti-oxidative effects of curcumin on immobilization-induced oxidative stress in rat brain, liver and kidney. *Biomed. Pharmacother.* **2017**, *87*, 223–229. [CrossRef] [PubMed]
49. Ansari, M.A.; Roberts, K.N.; Scheff, S.W. Oxidative stress and modification of synaptic proteins in hippocampus after traumatic brain injury. *Free. Radic. Biol. Med.* **2008**, *45*, 443–452. [CrossRef] [PubMed]
50. Limon-Pacheco, J.; Gonsebatt, M.E. The role of antioxidants and antioxidant-related enzymes in protective responses to environmentally induced oxidative stress. *Mutat. Res./Genet. Toxicol. Environ. Mutagenesis* **2009**, *674*, 137–147. [CrossRef] [PubMed]
51. Valko, M.; Rhodes, C.J.; Moncol, J.; Izakovic, M.; Mazur, M. Free radicals, metals and antioxidants in oxidative stress-induced cancer. *Chem.-Biol. Interact.* **2006**, *160*, 1–40. [CrossRef] [PubMed]
52. Girindrababu, V.J.; Krishnaswamy, K.; Puttasiddiah, R.; Farhath, K. Patulin induced oxidative stress mediated apoptotic damage in mice, and its modulation by green tea leaves. *J. Clin. Exp. Hepatol.* **2017**, *7*, 127–134.
53. He, M.D.; Xu, S.C.; Lu, Y.H.; Li, L.; Zhong, M.; Zhang, Y.W.; Wang, Y.; Li, M.; Yang, J.; Zhang, G.B.; et al. L-carnitine protects against nickel-induced neurotoxicity by mitochondrial function in Neuro-2a cells. *Toxicol. Appl. Pharmacol.* **2011**, *253*, 38–44. [CrossRef] [PubMed]

toxins

MDPI

Article

Protective Effects of the Hydroethanolic Extract of *Fridericia chica* on Undifferentiated Human Neuroblastoma Cells Exposed to α-Zearalenol (α-ZEL) and β-Zearalenol (β-ZEL)

Neda Alvarez-Ortega [1,2], Karina Caballero-Gallardo [1,2], María Taboada-Alquerque [1], Jackeline Franco [3], Elena E. Stashenko [4], Cristina Juan [5], Ana Juan-García [5,*] and Jesus Olivero-Verbel [1,*]

1 Environmental and Computational Chemistry Group, School of Pharmaceutical Sciences, Zaragocilla Campus, University of Cartagena, Cartagena 130014, Colombia; nalvarezo@unicartagena.edu.co (N.A.-O.); kcaballerog@unicartagena.edu.co (K.C.-G.); mtaboadaa@unicartagena.edu.co (M.T.-A.)
2 Functional Toxicology Group, School of Pharmaceutical Sciences, Zaragocilla Campus, University of Cartagena, Cartagena 130014, Colombia
3 Metabolite Profiling Facility, Bindley Bioscience Center, Purdue University, 1203 W State St., West Lafayette, IN 47907, USA; francoj@purdue.edu
4 Center for Chromatography and Mass Spectrometry CROM-MASS, Research Center for Biomolecules CIBIMOL, School of Chemistry, Universidad Industrial de Santander, Bucaramanga 680006, Colombia; elena@tucan.uis.edu.co
5 Facultat de Farmàcia—Avda, Universitat de València, Vicent Andrés Estellés, s/n, 46100 València, Spain; crisjua3@uv.es
* Correspondence: ana.juan@uv.es (A.J.-G.); joliverov@unicartagena.edu.co (J.O.-V.)

Citation: Alvarez-Ortega, N.; Caballero-Gallardo, K.; Taboada-Alquerque, M.; Franco, J.; Stashenko, E.E.; Juan, C.; Juan-García, A.; Olivero-Verbel, J. Protective Effects of the Hydroethanolic Extract of *Fridericia chica* on Undifferentiated Human Neuroblastoma Cells Exposed to α-Zearalenol (α-ZEL) and β-Zearalenol (β-ZEL). *Toxins* **2021**, *13*, 748. https://doi.org/10.3390/toxins13110748

Received: 28 September 2021
Accepted: 19 October 2021
Published: 22 October 2021

Abstract: *Fridericia chica* (Bignoniaceae) is a traditional medicinal plant. The aim of this research was to determine the protective effects of the hydroethanolic extract from the *F. chica* leaves (HEFc) against the cytotoxicity of zearalenone (α-ZEL) and β-ZEL on SH-SY5Y cells. Free radical scavenging activity of HEFc was evaluated using the DPPH method. The cytotoxicity of both zearalenone metabolites and HEFc was examined using MTT test, as was the cytoprotective effects of the HEFc on cells treated with these mycotoxins. The chemical composition of HEFc was determined using UPLC-QTOF-MS/MS. HEFc elicited good DPPH radical scavenging activity following a concentration-dependent relationship. Cells exposed to α-ZEL exhibited a viability <50% after 48 h of treatment (25 and 50 μM), while those exposed to β-ZEL showed viability <50% (100 μM) and <25% (25-100 μM) after 24 and 48 h of exposure, respectively. HEFc showed a significant increase in cell viability after exposure to α-ZEL (25 and 50 μM) and β-ZEL (6–100 μM) ($p < 0.05$). UPLC-QTOF-MS/MS analyses allowed the identification of 10 phytochemical components in the HEFc. In short, the hydroethanolic extract of *F. chica* grown in Colombian Caribbean can protect against the effects of mycotoxins and it is a valuable source of compounds with antioxidant properties.

Keywords: cytotoxicity; mycotoxins; *Fridericia chica*; protection; extracts

Key Contribution: The hydroethanolic extract of the leaves of *F. chica* is capable of suppressing the cellular proliferation of human neuroblastoma cells induced by the metabolites of zearalenone.

1. Introduction

Colombia is one of the countries in the world that is considered to be mega-diverse due to its various ecosystems [1]. Bioprospecting has always been a central activity in human development; it is defined as the evaluation of biological material in order to search for new valuable products, and involves the application of advanced technologies for the development of new pharmaceutical, agrochemical products, cosmetics, flavorings, fragrances, industrial enzymes, and other products from biota. The Colombian Caribbean flora offers countless possibilities; an example of this is the *Fridericia chica* (Bonpl.) L.G.

Lohmann, Bignoniaceae, which is known as "*limpia diente*" and grows frequently in tropical regions in South America and Africa.

Fridericia chica (synonym *Arrabidaea chica*) leaves are usually employed as a red dye, in fact, it has traditionally been used by Sinu artisans of the Colombian Caribbean to make the *vueltiao* hat [2]. The red color obtained from *F. chica* comes from anthocyanidins, a class of phenolic compounds with known antioxidant properties [3], and extracts from the plant have shown pharmacological activities linked to beneficial health effects, including anti-inflammatory [4], antiproliferative [5], wound healing [6], antispasmodic [7], photoprotective [8], and leishmanicidal [9] activities. Many of these properties are attributed to different flavonoid compounds reported in *F. chica*, such as isoscutellarein, 6-hydroxyluteolin, hispidulin, scutellarein, luteolin, apigenin, and hispidulin [10], this last a potential compound for neuroinflammation inhibition [11].

Exposure to environmental toxins is key in the development of neuropathological diseases [12]. Within these pollutants, the mycotoxins are toxic secondary metabolites of fungi and are commonly produced in stored agricultural products [13]. They represent a risk for human and animal health due to their potential to cause disease and death in organisms that are exposed to them [14]. Zearalenone (ZEA) metabolites such as α-zearalenol (α-ZEL) and β-zearalenol (β-ZEL) are produced by the genus *Fusarium* and infect mainly cereals and other plant products [15]. The toxicity of these metabolites is related to their ability to bind to the estrogen receptor and their ability to modify estrogen metabolism [16]. In addition, these metabolites are known oxidative stressors [15,17]. In fact, it has been reported that α-ZEL can induce proliferation in different human cell lines such as MCF7 [18] and granulosa cells [19].

The human neuroblastoma dopaminergic neuronal cell line SH-SY5Y has been used as an in vitro model for numerous neurotoxicity and cytoprotection experiments. The aim of this work was to investigate the potential of a hydroethanolic extract of *F. chica* (HEFc) grown in northern Colombia to mitigate the effects of α-ZEL and β-ZEL in undifferentiated human neuroblastoma, as well as to characterize its constituents.

2. Results

2.1. DPPH Radical Scavenging

The results of the free radical scavenging activity of the HEFc are presented in Figure 1. The HEFc quenched DPPH free radicals in a concentration-dependent manner. The antioxidant assay produced values in the range of 21 ± 4.5–430 ± 27.8 μM Trolox. At 16 μg/mL, HEFc showed a low DPPH inhibition of 6%, and at 1000 μg/mL, it resulted in 65% DPPH inhibition. In addition, the IC_{50} value was found to be 709 μg/mL.

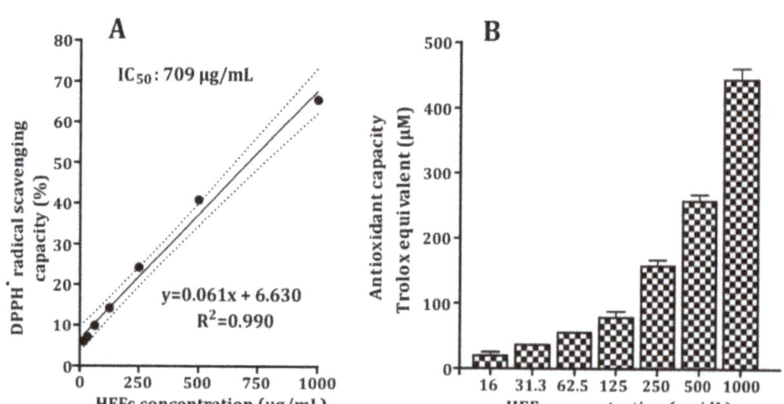

Figure 1. DPPH radical scavenging capacity (**A**) and Antioxidant capacity, Trolox equivalent (μM) (**B**).

2.2. Cytotoxicity of HEFc and Mycotoxins on SH-SY5Y Cells

The cytotoxic activity of the HEFc over 24 and 48 h in undifferentiated human neuroblastoma cells is presented in Figure 2. The HEFc was found to significantly reduce cell viability in SH-SY5Y cells. It was determined that exposure to a concentration of 16 µg/mL for 24 to 48 h periods did not affect viability (Figure 2). The corresponding IC_{50} values for SH-SY5Y cells at 24 and 48 h were 61.2 µg/mL (45 to 83 µg/mL) and 53.8 µg/mL (24.0 to 116.6 µg/mL), respectively (Table 1).

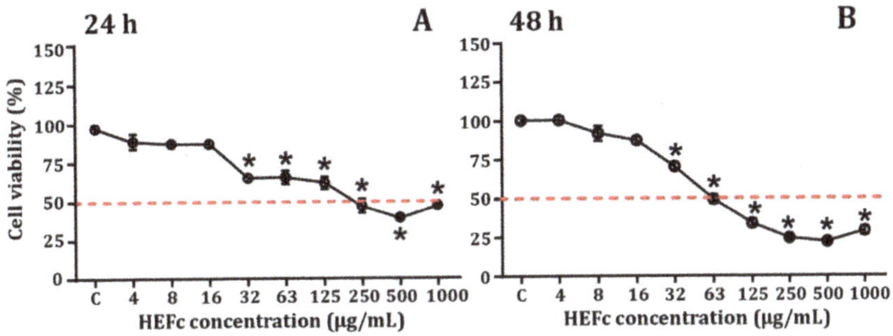

Figure 2. Efect of HEFc on SH5YSY viability. (**A**) Cytotoxicity effects in undifferentiated human neuroblastoma cells exposed to HEFc for 24 h, (**B**) Cytotoxicity effects in undifferentiated human neuroblastoma cells exposed to HEFc for 48 h. * $p < 0.05$, representing a significant difference compared to the control. Data are mean ± SEM ($n = 3$), one-way ANOVA post Sidak's multiple comparisons test.

Table 1. Medium inhibitory concentration IC_{50} in SH-SY5Y cells exposed to zearalenone metabolite and HEFc over 24 and 48 h periods.

Treatment	IC_{50} (CI_{95}) 24 h	IC_{50} (CI_{95}) 48 h
HEFc	61.2 µg/mL (45–83)	53.8 µg/mL (24.0–116.6)
α-ZEL	>50 µM	17.9 µM (10.4–32.4)
β-ZEL	>100 µM	10.5 µM (7.1–15.7)

The cytotoxicity of α-ZEL and β-ZEL in undifferentiated human neuroblastoma cells are displayed in Figure 3. Cell proliferation increased at low concentrations (0.4, 0.8 and 1.6 µM), after 24 and 48 h of exposure; β-ZEL also caused a similar behavior, but only during the first 24 h and the IC_{50} could not be calculated after 24 h of treatment; while at 48 h of treatment, the IC_{50} values were 17.9 µM (CI_{95}: 10.4 to 32.4 µM) for α-ZEL and 10.5 µM (CI_{95}: 7.1 to 15.7 µM) for β-ZEL (Table 1).

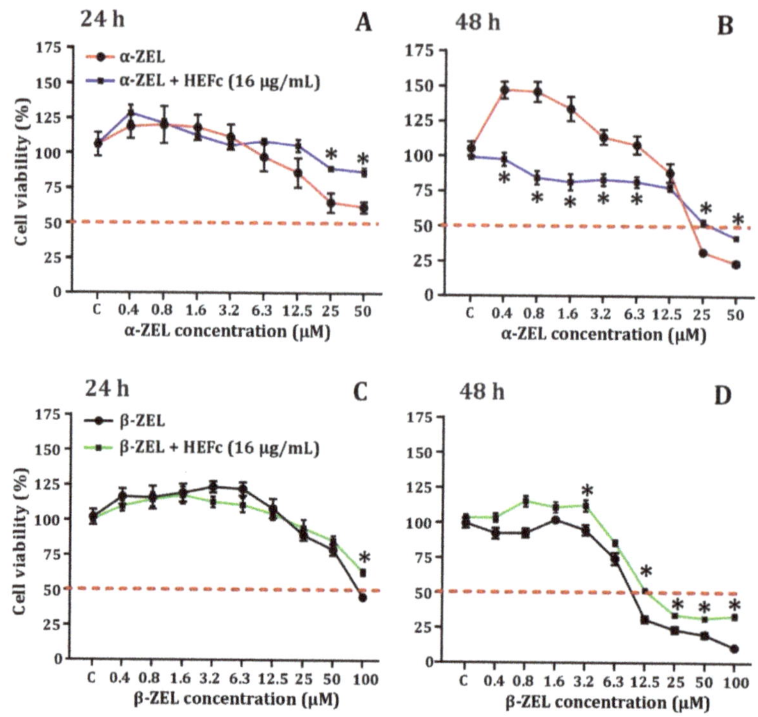

Figure 3. Cell viabillity in SH-SY5Y exposed to α-ZEL/α-ZEL/HEFc (**A**,**B**) or β-ZEL/β-ZEL/HEFc (**C**,**D**) over 24 (**A**,**C**) and 48 h (**B**,**D**). * $p < 0.05$, significant difference compared to the correspondiong metabolite tested alone. Data are mean ± SEM ($n = 3$). Multiple t-tests.

2.3. Cytoprotective Effects of HEFc against Zearalenone Metabolites

The protective effects of HEFc against cytotoxicity in SH-SY5Y cells induced by α-ZEL and β-ZEL after 24 and 48 h exposures, are shown in Figure 3. The results showed that when SH-SY5Y cells were simultaneously treated with α-ZEL (25 and 50 μM) and HEFc (16 μg/mL), the extract exhibited significant cytoprotection (24–25%, $p < 0.05$) after the first 24 h of incubation. HEFc (16 μg/mL) significantly increased in cell proliferation in samples with low concentrations of α-ZEL (0.4, 0.8, 1.6, 3.2, and 6.3 μM, $p < 0.05$) after treatment for 48 h. At the highest α-ZEL concentrations tested (25–50 μM), simultaneous co-exposure with the extract was able to significantly increase cell viability ($p < 0.05$) by 18–21%.

The treatment with HEFc (16 μg/mL) in undifferentiated human neuroblastoma cells exposed for 24 h with β-ZEL only displayed some protection in viability (~20%) at the highest tested mycotoxin concentration (100 μM) tested. After prolonged exposure (48 h), HEFc was able to significantly recover cell viability at an extent similar to that observed after 24 h (~20%); however, the concentration range displaying the effect was greater (12.5–100 μM) (Figure 3).

2.4. UPLC-QTOF-MS/MS Analysis

The results of ultra-performance liquid chromatography/quadrupole time-of-flight mass spectrometry (UPLC-QTOF-MS/MS) analysis of the HEFc are shown in Figure 4 and Table 2. The identified compounds belong to various classes, including flavone glycosides (Vicenin-2, 6-hydroxyluteolin 7-rhamnoside and Scutellarein-O-glucuronide) and flavones (Nepetin, Pectolinarigenin, Hispidulin, Apigenin, 4′,6,7-trihydroxy-5-methoxyflavone, Thevetiaflavone and Acacetin), among others. The extracted ion chromatogram (Rt: 3–11 min, EIC) for the HEFc obtained by UPLC-QTOF-MS/MS in positive ion mode, the exact mass characteristics for positive ions of identified chemicals, and fragment spectrum results (MS/MS) are presented in Supplementary Material.

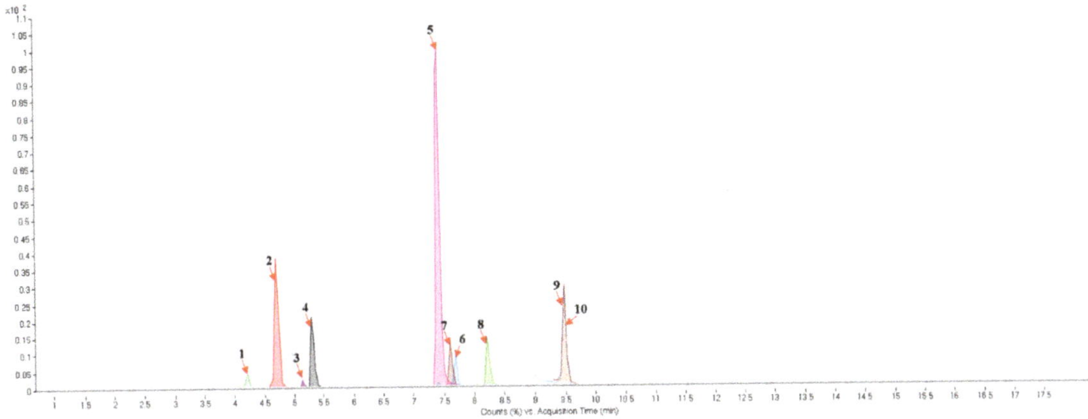

Figure 4. Chromatograms of *Fridericia chica*. Extract chromatogram Rt: 3–11 min. (extracted ion chromatogram, EIC), obtained by UPLC-MS/MS in positive ion mode. Numbers correspond to (1) vicenin-2; (2) 6-hydroxyluteolin 7-rhamnoside; (3) scutellarein-O-glucuronide; (4) nepetin; (5) pectolinarigenin; (6) hispidulin; (7) apigenin; (8) 5-O-methylscutellarein; (9) thevetiaflavone and (10) acacetin.

Table 2. Top results of UPLC-QTOF-MS/MS analysis.

No. Figure 4	RT (min)	Tentative Annotation	Structure	Formula	Ion	Experimental Mass	Calculated Mass	Δ ppm
1	4.204	Vicenin-2	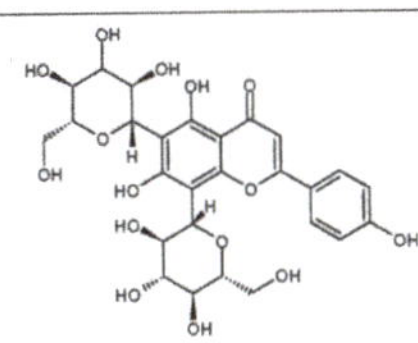	$C_{27}H_{30}O_{15}$	$[M+H]^+$	594.1571	594.15847	−2.30
2	4.732	6-hydroxyluteolin 7-rhamnoside	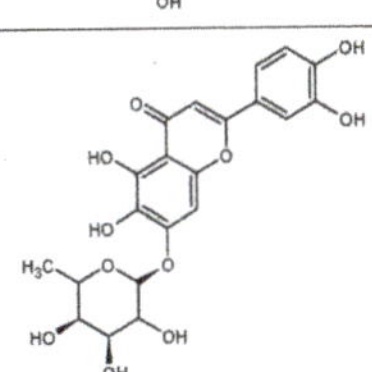	$C_{21}H_{20}O_{11}$	$[M+H]^+$	448.10071	448.10056	0.33
3	5.158	Scutellarein-O-glucuronide	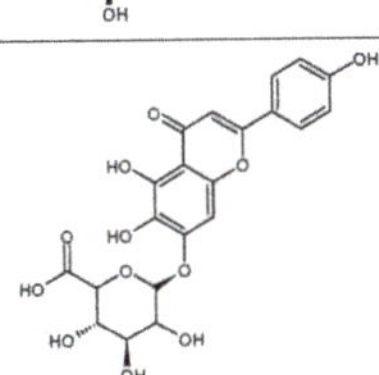	$C_{21}H_{18}O_{12}$	$[M+H]^+$	462.07918	462.07983	−1.40

Table 2. *Cont.*

No. Figure 4	RT (min)	Tentative Annotation	Structure	Formula	Ion	Experimental Mass	Calculated Mass	Δ ppm
4	5.355	Nepetin	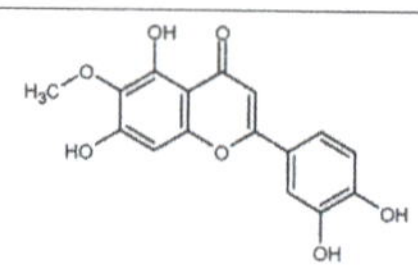	$C_{16}H_{12}O_7$	$[M+H]^+$	316.05831	316.0583	0.03
5	7.443	Pectolinarigenin	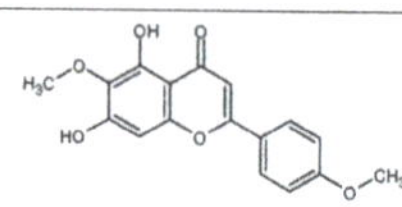	$C_{17}H_{14}O_6$	$[M+H]^+$	314.07919	314.07904	0.47
6	7.578	Hispidulin	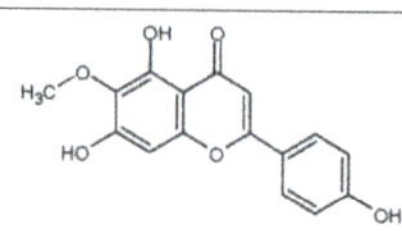	$C_{16}H_{12}O_6$	$[M+H]^+$	300.06317	300.06339	−0.73
7	7.631	Apigenin	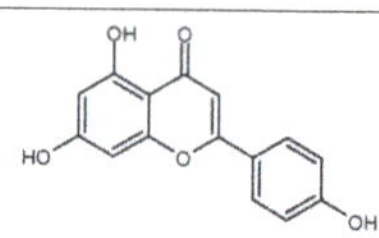	$C_{15}H_{10}O_5$	$[M+H]^+$	270.05259	270.05282	−0.85
8	8.254	5-O-methylscutellarein	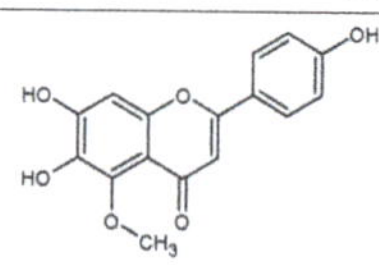	$C_{16}H_{12}O_6$	$[M+H]^+$	300.06367	300.06339	0.93
9	9.331	Thevetiaflavone	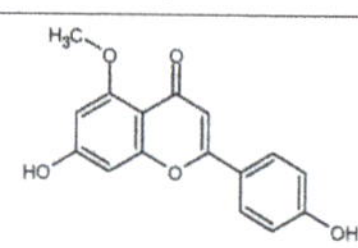	$C_{16}H_{12}O_5$	$[M+H]^+$	284.0685	284.06847	0.10
10	9.572	Acacetin	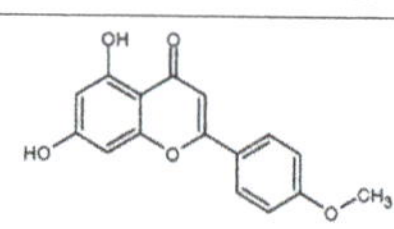	$C_{16}H_{12}O_5$	$[M+H]^+$	284.06872	284.06847	0.88

3. Discussion

This study evaluated the free radical scavenging activity of a hydroethanolic extract of *F. chica* (HEFc) obtained from fallen leaves collected from trees grown in the Colombian Caribbean, the protective effects of the extract on zearalenone metabolites-induced citotoxicity in SH-SY5Y cells, and its chemical composition.

F. chica is a plant widely known for its anti-inflammatory [4,5], antibacterial [20,21] healing actions [6], and antioxidant properties [22,23]. However scientific studies of its properties from fallen leaves of plants cultivated in Colombia are scarce. The DPPH scavenging activity was dependent concentration of HEFc. The IC_{50} of the extract was 709 µg/mL, a value much than that reported in extracts obtained from leaves of the same species from Brazil (IC_{50} of 13.5 µg/mL) [10] or Argentina (57.84 µg/mL) [23]. These differences may result from distinct climatic conditions or soil properties of the sites where they have been collected, but this is unclear, as this is the first study that reports the DPPH antiradical activity of HEFc in the Colombian Caribbean. The antioxidant properties of this plant can be explained by the presence of flavonoids, alkaloids, and phenolic compounds found in phytochemical screening also reported by other authors [24,25].

Although some biological properties of *F. chica* have been previously reported [23,25]; the present research is the first to evaluate the response on the viability of undifferentiated human neuroblastoma cells exposed to HEFc from the Colombian Caribbean. Interestingly, cell proliferation was not observed under any of the concentrations evaluated. The IC_{50} found here after 24 and 48 h treatment (61.2 and 53.8 μg/mL), were moderately higher than the IC_{50} values (<30 μg/mL) suggested as a criteria to extract promising agents for anticancer drug development [26]. However, a similar extract has shown good activities in other cancer cell lines, such as HL60 (IC_{50}, 26.9 μg/mL), and Jurkat cells (IC_{50}, 27.9 μg/mL) [5]. Notably, the extract has been reported to show growth stimulation in several cell lines, including fibroblasts [6], NIH-3T3 cells [27], and CHO-K1 cells [20].

Zearalenone metabolites at high concentrations produce an increase in cell death after the first 24 h of treatment, finally reaching values below the IC_{50} after 48 h of exposure. These results are similar to those obtained by other authors [16,28]. However, there are differences with respect to the concentrations required to reach IC_{50} values after 48 h of exposure. In the present study, it was found that a concentration of 17.9 μM (α-ZEL) and 10.5 μM (β-ZEL) was required to reach the IC_{50} values in SH-SY5Y cells, while other types of cell lines appear less sensitive (e.g., IC_{50} of 32 and 55 μM for α-ZEL and β-ZEL, respectively in CHO-K1 cells) [28]. Interestingly, α-ZEL induced and increase in cell proliferation at low concentrations. These results are consistent with those reported by other authors [19]. It is known that cell proliferation is an essential event in various physiological processes, such as tissue generation, but also in various pathophysiological events such as cancer formation [29]. In this last case, the evidence suggests ZEA metabolites stimulate cell proliferation and therefore may promote cancer in different cells [30].

The brain is also a target for estrogens and phytoestrogens, such as ZEA or its metabolites, as they are known to cross the blood-brain barrier [31]. Although data are limited and their role of mycotoxins in neurodegerative diseases is not yet understood, a recent study showed these molecules alter the expression of dopaminergic genes in SH5YSY cells [32]. In this work, the HEFc exerted an antiproliferative effect on neuronal cells, suggesting it has propective properties on ZEA metabolites-induced cell proliferation.

The characterization of HEFc showed that some of the tentatively identified compounds has been reported to have potential anti-osteoporotic, anti-inflammatory, antiplatelet, anticonvulsant, and anticancer [33,34]. Flavone compounds such as nepetin, pectolinarigenin, apigenin, taxifolin, hispidulin, thevetiaflavone, and acacetin present in the HEFc have been reported as potential neuroprotective agents for their ability to suppress neuronal apoptosis, oxidative stress and inflammation by attenuating levels of malondialdehyde, lactate dehydrogenase, Bax, caspase-3, TNF-α, IL-1β and IL-6, and increasing levels of Bcl-2 and SOD [35–40].

4. Conclusions

The present study demonstrates that hydroethanolic extract from the *F. chica* leaves exerts significant protective effects against mycotoxin-induced cytotoxicity in undifferentiated human neuroblastoma cells. This may be due to the presence of compounds with antioxidant properties present in the extract. These results may contribute to new approaches for treatment against the effects of mycotoxins.

5. Materials and Methods

5.1. Plant Material and Extraction

The fallen leaves of *F. chica* (Bonpl.) L.G. Lohmann, Bignoniaceae were collected between January and December 2018 in the municipality of Sincelejo (9°14′20″ N-75°25′17″ W), department of Sucre (Sincelejo, Colombia). The specimens were identified by Pedro Alvarez Perez in the Herbarium of the University of Sucre (Sincelejo, Colombia), and a voucher specimen was deposited and registered in this herbarium (004537). Leaves of *F. chica* were dried at room temperature. Subsequently, dried leaves were cut and powder (30 g) was mixed with (1:10, *w*/*v*) with 70% ethanol in water for 4 h at 200 rpm using an automated Shaker. The extract was filtered and the solvent evaporated in an oven at 70 °C under a hood. The extract was a black reddish solid, and was stored at −20 °C until analysis.

5.2. DPPH Radical-Scavenging Method

DPPH radical scavenging assay was determined using the kit from Bioquochem (Llanera, Asturias, Spain). Initially, a stock solution at 100,000 μg/mL of the hydroethanolic extract from the HEFc with MilliQ water was prepared. All extract samples were tested in duplicate at seven different concentrations (1000, 500, 250, 125, 62.5, 32.5, and 16 μg/mL), then, sample extracts (20 μL) were reacted with 200 μL of the DPPH solution in the dark at room temperature. The absorbance at 517 nm was performed in the Varioskan™ LUX Multimode Microplate Reader (Thermo Fisher Scientific, Inc., Waltham, MA, USA). The percentage of inhibition of the radical DPPH•+ for each standard point was obtained with the following formula:

$$\% \text{ Inhibition} = [1 - (\text{Abs Sn}/\text{Abs S1})] \times 100 \tag{1}$$

where Abs S1 is the DPPH•+ radical absorption without inhibition and Abs Sn is the DPPH•+ radical absorption of the correspondent standard.

A Trolox^R standard curve was used to determine TEAC (Trolox Equivalent Antioxidant Capacity) of the tested concentrations of extracts. The calculation was performed using the following formula:

$$\text{TEAC } (\mu\text{M}) = \% \text{ inhibition} - \text{intercept}/\text{slope} \tag{2}$$

5.3. Cellular Exposure to Mycotoxins and HEFc

SH-SY5Y NB cells (ATCC® CRL-2266™) were maintained following the methods previously reported [41]. The cells were incubated in 96-well plates (2×10^6 cells/plate) for 24 h. Briefly, the cells were exposed to 100 μL of medium containing different concentrations of α-ZEL (0.4 to 50 μM), β-ZEL (0.4 to 100 μM) and HEFc (4 to 1000 μg/mL) individually for 24 and 48 h. For all cases, the viability was examined using MTT test (Section 5.5), and absorbance was determined using a spectrophotometer at 570 nm (VICTOR x5 Multimode Microplate, PerkinElmer's, Waltham, MA, USA). Three experiments were carried out with four replicates each. Concentration-effect curves were built to obtain IC_{50} using Prism 6.0 (GraphPad Software Inc.).

5.4. Cytoprotective Effects of HEFc against ZEN Metabolites

SH-SY5Y cells incubated in 96-well plates (2×10^6 cells/plate) for 24 h at 37 °C in a 5% CO_2 atmosphere and these were exposed to two independent treatment combinations. The first consisted of a mixture of HEFc (16 μg/mL) with eigth concentrations of α-ZEL (from 0.4 to 50 μM, 1: 2 dilutions), for the second, the combination of HEFc (16 μg/mL) and nine concentrations of ZEN metabolites (0.4 to 100 μM, 1: 2 dilutions) for 24 and 48 h were tested. The plates were incubated for 24 and 48 h at 37 °C, in a 5% CO_2 atmosphere and the viability was examined by MTT assay (Section 5.5). Three experiments were carried out with four replicates each.

5.5. MTT Assay

Viability was examined using the MTT assay as previously reported [42,43]. Briefly, medium with treatments (HEFc, mycotoxins, and mycotoxins/HEFc), as previously described, was removed and each well received 200 μL of medium with 50 μL of MTT solution (5 mg/mL). After an incubation period of 4 h at 37 °C, the MTT containing media was discarded and 200 μL of DMSO and 25 μL of Sorensen's solution were added to each well prior to reading optical density at 570 nm, using a VICTOR x5 multimode plate reader (PerkinElmer, Waltham, MA, USA). All experiments were performed in three independent experiments with four replicates for each treatment.

5.6. Analysis of the HEFc by UPLC-QTOF-MS/MS

The HEFc was diluted 1/10 in 50% acetonitrile mixed with 0.1% HCOOH. Samples were sonicated for 5 min and then centrifuged at 16,000 *g* (8 min). The supernatants were transferred to auto sampler vials. Ultra-performance liquid chromatography/quadrupole time-of-flight mass spectrometry (UPLC-QTOF-MS/MS) was utilized to separate and characterize the components. Chromatographic separation was performed using an Agilent 1290 UPLC system (Agilent Technologies, Palo Alto, CA, USA) employing a YMC Carotenoid column (3 μm particle size, 2.0 × 150 mm) (YMC America Allenton, PA, USA) with a mobile phase flow rate of 0.3 mL/min, where the mobile phase A and B were 0.1% HCOOH acid in ddH_2O and acetonitrile, respectively. Starting conditions were 95:5 A: B, held for 1 min, followed by a linear gradient to 5:95 at 12 min, with a hold to 15 min. Column re-equilibration was accomplished by returning to 95:5 A:B at 16 min and holding until 21 min. The mass analysis was acquired with an Agilent 6545 Q-TOF MS (Agilent Technologies, Santa Clara, CA, USA) with ESI capillary voltage +3.5 kV, N_2 gas temperature 320 °C, drying gas flow rate 8.0 L/min, nebulizer gas pressure 35 psig, fragmentor voltage 135 V, skimmer 65 V, and OCT RF 750 V. Mass spectral data were gathered in profile mode. Mass accuracy was enhanced by infusing Agilent Reference Mass Correction Solution (G1969-85001). The instrument was operated from 100 to 1200 *m/z* at a scan rate of 2 spectra/s. MS data scans were obtained using Agilent MassHunter Acquisition software (v. B.06). MS/MS was achieved in a data-dependent acquisition mode on composite samples. Peak deconvolution and integration was executed using Agilent ProFinder (v. B.06). Peak annotations were carried out with METLIN (metlin.scripps.edu) metabolite databases, with a mass error around 1 ppm. Identifications were supported by MS/MS spectra comparisons.

5.7. Statistical Analysis

The data are presented as mean±SEM and analyzed statistically by GraphPad Prism 8.0 (GraphPad Prisma Software, Inc., San Diego, CA, USA). The IC_{50} values of HEFc and ZEN metabolites were analyzed using a non-linear sigmoid curve fit. Statistical comparison was performed using one-way analysis of variance (ANOVA) with Sidak's multiple comparisons test. In addition, multiple Student *t*-test was conducted to investigate different mean cell viabillity between mixture (extract and metabolite) and metabolite alone. Results were considered significant at $p < 0.05$.

Supplementary Materials: The following are available online at https://www.mdpi.com/article/10.3390/toxins13110748/s1, Table S1: Exact mass characteristic positive ions of compounds identified by UPLC-QTOF-MS/MS in *Fridericia chica* extract, Figure S1: Compound fragment spectrum results (MS/MS).

Author Contributions: Data curation, N.A.-O.; Formal analysis, N.A.-O., K.C.-G., J.O.-V.; Funding acquisition, J.O.-V., E.E.S., C.J., A.J.-G.; Investigation, J.O.-V., K.C.-G., A.J.-G., E.E.S.; Methodology, N.A.-O., J.F., M.T.-A.; Supervision, J.O.-V., K.C.-G., C.J., A.J.-G.; Writing—original draft, N.A.-O., K.C.-G., J.O.-V.; Writing—review & editing, J.O.-V., K.C.-G., A.J.-G. All authors have read and agreed to the published version of the manuscript.

Funding: The authors thank the Ministry of Science, Technology and Innovation (Minciencias), the Ministry of Education, the Ministry of Industry, Commerce and Tourism, and ICETEX, Programme Ecosistema Científico-Colombia Científica, from the Francisco José de Caldas Fund, Grant RC-FP44842-212-2018; Spanish Ministry of Science and Innovation PID 2019-108070RB-I00ALI; The Ministry of Environment and Sustainable Development of Colombia supported the Universidad Industrial de Santander through access permits to genetic resources and derivatives for bioprospecting (Contract No. 270-2019); Minciencias, Sistema General de Regalías de Colombia (BPIN 2020000100093, Gobernación de Bolívar).

Acknowledgments: The authors thank Minciencias, Grant 727-2015, FJC Fund FP44842-169-2019 and Generalitat Valenciana GV 2020/020.

Conflicts of Interest: The authors declare no conflict of interest.

Abbreviations

CI_{95}	95% confidence interval
IC_{50}	inhibitory concentration
α-ZEL	α-zearalenol
β-ZEL	β-zearalenol
HEFc	hydroethanolic extract from *F. chica* leaves
SEM	standard error of mean
ZEN	zearalenone

References

1. Humboldt. Instituto de Investigación de Recursos Biológicos Alexander von Humboldt. 2020. Available online: www.humboldt.org.co (accessed on 19 October 2021).
2. Castillo, B.O.D.; Jiménez, Y.M.; Nieves, M.F.F.; Peláez, J.A.M. Composición química de aceites esenciales de hojas de *Fridericia florida* DC. y *Fridericia chica* (Bonpl.). *Rev. Fac. Cienc. Bás.* **2019**, *15*, 63–70. [CrossRef]
3. do Amaral, R.R.; Santos, A.A.; Saravia, A.; Botas, G.; Cruz, R.A.; Fernandes, C.P.; Rocha, L.; Boylan, F. Biological activities of *Arrabidaea chica* (Bonpl.) B. Verl. leaves. *Lat. Am. J. Pharm.* **2012**, *31*, 451–455.
4. Lima, J.C.; de Oliveira, R.G.; Silva, V.C.; de Sousa Jr, P.T.; Violante, I.M.; Macho, A.; Martins, D.T.d.O. Anti-inflammatory activity of 4′, 6, 7-trihydroxy-5-methoxyflavone from *Fridericia chica* (Bonpl.) LG Lohmann. *Nat. Prod. Res.* **2020**, *34*, 726–730. [CrossRef] [PubMed]
5. Michel, A.F.; Melo, M.M.; Campos, P.P.; Oliveira, M.S.; Oliveira, F.A.; Cassali, G.D.; Ferraz, V.P.; Cota, B.B.; Andrade, S.P.; Souza-Fagundes, E.M. Evaluation of anti-inflammatory, antiangiogenic and antiproliferative activities of *Arrabidaea chica* crude extracts. *J. Ethnopharmacol.* **2015**, *165*, 29–38. [CrossRef] [PubMed]
6. Jorge, M.P.; Madjarof, C.; Ruiz, A.L.T.G.; Fernandes, A.T.; Rodrigues, R.A.F.; de Oliveira Sousa, I.M.; Foglio, M.A.; de Carvalho, J.E. Evaluation of wound healing properties of *Arrabidaea chica* Verlot extract. *J. Ethnopharmacol.* **2008**, *118*, 361–366. [CrossRef]
7. Monteiro, F.D.S.; Costa, J.R.D.S.; Martins, L.J.A.; Rocha, C.Q.; Borges, A.C.R.; Borges, M.O.D.R. Hydroalcoholic extract of leaves of *Arrabidaea brachypoda* (DC.) Bureau present antispasmodic activity mediated through calcium influx blockage. *Rev. Ciênc. Farm. Básica Apl.* **2020**, *41*, 1–13. [CrossRef]
8. Siraichi, J.T.; Pedrochi, F.; Natali, M.R.; Ueda-Nakamura, T.; Filho, B.P.; Bento, A.C.; Baesso, M.L.; Nakamura, C.V. Ultraviolet (UVB and UVA) photoprotector activity and percutaneous penetration of extracts obtained from *Arrabidaea chica*. *Appl. Spectrosc.* **2013**, *67*, 1179–1184. [CrossRef]
9. Cortez de Sá, J.; Almeida-Souza, F.; Mondêgo-Oliveira, R.; Oliveira Idos, S.; Lamarck, L.; Magalhães Ide, F.; Ataídes-Lima, A.F.; Ferreira Hda, S.; Abreu-Silva, A.L. Leishmanicidal, cytotoxicity and wound healing potential of *Arrabidaea chica* Verlot. *BMC Complement. Altern. Med.* **2015**, *16*, 1–11. [CrossRef]
10. Siraichi, J.T.; Felipe, D.F.; Brambilla, L.Z.; Gatto, M.J.; Terra, V.A.; Cecchini, A.L.; Cortez, L.E.; Rodrigues-Filho, E.; Cortez, D.A. Antioxidant capacity of the leaf extract obtained from *Arrabidaea chica* cultivated in Southern Brazil. *PLoS ONE* **2013**, *8*, e72733. [CrossRef]
11. Yu, C.I.; Cheng, C.I.; Kang, Y.F.; Chang, P.C.; Lin, I.P.; Kuo, Y.H.; Jhou, A.J.; Lin, M.Y.; Chen, C.Y.; Lee, C.H. Hispidulin inhibits neuroinflammation in lipopolysaccharide-activated BV2 microglia and attenuates the activation of Akt, NF-κB, and STAT3 pathway. *Neurotox Res.* **2020**, *38*, 163–174. [CrossRef]
12. Juan-García, A.; Juan, C.; Bind, M.A.; Engert, F. Study of locomotion response and development in zebrafish (Danio rerio) embryos and larvae exposed to enniatin A, enniatin B, and beauvericin. *Sci. Total Environ.* **2021**, *777*, 146075. [CrossRef] [PubMed]
13. Wang, H.W.; Wang, J.Q.; Zheng, B.Q.; Li, S.L.; Zhang, Y.D.; Li, F.D.; Zheng, N. Cytotoxicity induced by ochratoxin A, zearalenone, and α-zearalenol: Effects of individual and combined treatment. *Food Chem. Toxicol.* **2014**, *71*, 217–224. [CrossRef] [PubMed]
14. Agahi, F.; Juan, C.; Font, G.; Ana Juan-García, A. In silico methods for metabolomic and toxicity prediction of zearalenone, α-zearalenone and β-zearalenone. *Food Chem. Toxicol.* **2020**, *146*, 111818. [CrossRef]
15. Tatay, E.; Espin, S.; Garcia-Fernandez, A.-J.; Ruiz, M.-J. Oxidative damage and disturbance of antioxidant capacity by zearalenone and its metabolites in human cells. *Toxicol. Vitr.* **2017**, *45*, 334–339. [CrossRef] [PubMed]
16. Tiemann, U.; Viergutz, T.; Jonas, L.; Schneider, F. Influence of the mycotoxins α-and β-zearalenol and deoxynivalenol on the cell cycle of cultured porcine endometrial cells. *Reprod. Toxicol.* **2003**, *17*, 209–218. [CrossRef]
17. Agahi, F.; Juan, C.; Font, G.; Juan-García, A. Neurotoxicity of zearalenone's metabolites and beauvericin mycotoxins via apoptosis and cell cycle disruption. *Toxicology* **2021**, *456*, 152784. [CrossRef]
18. Shier, W.T.; Shier, A.; Xie, W.; Mirocha, C. Structure-activity relationships for human estrogenic activity in zearalenone mycotoxins. *Toxicon* **2001**, *39*, 1435–1438. [CrossRef]
19. Pizzo, F.; Caloni, F.; Schutz, L.F.; Totty, M.L.; Spicer, L.J. Individual and combined effects of deoxynivalenol and α-zearalenol on cell proliferation and steroidogenesis of granulosa cells in cattle. *Environ. Toxicol. Pharmacol.* **2015**, *40*, 722–728. [CrossRef]
20. Violante, I.M.; Carollo, C.A.; Silva, L.I.; Oliveira, A.Q.; Pardinho, F.C.; Garcez, W.S.; Garcez, F.R.; Oliveira, R.G.d.; Arunachalam, K.; de Oliveira Martins, D.T. Cytotoxicity and antibacterial activity of scutellarein and carajurone-enriched fraction obtained from the hydroethanolic extract of the leaves of *Fridericia chica* (Bonpl.) LG Lohmann. *Nat. Prod. Res.* **2020**, 1–7. [CrossRef]

21. Mafioleti, L.; da Silva Junior, I.F.; Colodel, E.M.; Flach, A.; de Oliveira Martins, D.T. Evaluation of the toxicity and antimicrobial activity of hydroethanolic extract of *Arrabidaea chica* (Humb. & Bonpl.) B. Verl. *J. Ethnopharmacol.* **2013**, *150*, 576–582. [PubMed]
22. Martins, F.J.; Caneschi, C.A.; Vieira, J.L.; Barbosa, W.; Raposo, N.R. Antioxidant activity and potential photoprotective from amazon native flora extracts. *J. Photochem. Photobiol. B* **2016**, *161*, 34–39. [CrossRef] [PubMed]
23. Torres, C.A.; Pérez Zamora, C.M.; Nuñez, M.B.; Gonzalez, A.M. In vitro antioxidant, antilipoxygenase and antimicrobial activities of extracts from seven climbing plants belonging to the Bignoniaceae. *J. Integr. Med.* **2018**, *16*, 255–262. [CrossRef] [PubMed]
24. Paula, J.T.; Paviani, L.C.; Foglio, M.A.; Sousa, I.M.O.; Duarte, G.H.B.; Jorge, M.P.; Eberlin, M.N.; Cabral, F.A. Extraction of anthocyanins and luteolin from *Arrabidaea chica* by sequential extraction in fixed bed using supercritical CO_2, ethanol and water as solvents. *J. Supercrit. Fluids* **2014**, *86*, 100–107. [CrossRef]
25. Olivero-Verbel, J.; De la Parra-Guerra, A.; Caballero-Gallardo, K.; Sierra-Marquez, L.; Fuentes-Lopez, K.; Franco-Marmolejo, J.; Jannasch, A.S.; Sepulveda, M.S.; Stashenko, E. The aqueous extract of *Fridericia chica* grown in northern Colombia ameliorates toxicity induced by tergitol on *Caenorhabditis elegans*. *Comp. Biochem. Physiol. C Toxicol. Pharmacol.* **2021**, *244*, 109026. [CrossRef] [PubMed]
26. Suffness, M.; Pezzuto, J.M. *Methods in Plant Biochemistry: Assays for Bioactivity*; Academic Press: London, UK, 1990; pp. 71–133.
27. Salles, T.H.C.; Volpe-Zanutto, F.; de Oliveira Sousa, I.M.; Machado, D.; Zanatta, A.C.; Vilegas, W.; Lancellotti, M.; Foglio, M.A.; d'Ávila, M.A. Electrospun PCL-based nanofibers *Arrabidaea chica* Verlot-Pterodon pubescens Benth loaded: Synergic effect in fibroblast formation. *Biomed. Mater.* **2020**, *15*, 65001. [CrossRef]
28. Othmen, Z.O.-B.; El Golli, E.; Abid-Essefi, S.; Bacha, H. Cytotoxicity effects induced by zearalenone metabolites, α-zearalenol and β-zearalenol, on cultured Vero cells. *Toxicology* **2008**, *252*, 72–77. [CrossRef]
29. Pillay, D.; Chuturgoon, A.A.; Nevines, E.; Manickum, T.; Deppe, W.; Dutton, M.F. The quantitative analysis of zearalenone and its derivatives in plasma of patients with breast and cervical cancer. *Clin. Chem. Lab. Med.* **2002**, *40*, 946–951. [CrossRef]
30. Zheng, W.; Wang, B.; Li, X.; Wang, T.; Zou, H.; Gu, J.; Yuan, Y.; Liu, X.; Bai, J.; Bian, J.; et al. Zearalenone promotes cell proliferation or causes cell death? *Toxins* **2018**, *10*, 184. [CrossRef]
31. Gonkowski, S.; Obremski, K.; Calka, J. The influence of low doses of zearalenone on distribution of selected active substances in nerve fibers within the circular muscle layer of porcine ileum. *J. Mol. Neurosci.* **2015**, *56*, 878–886. [CrossRef]
32. Venkataramana, M.; Chandra Nayaka, S.; Anand, T.; Rajesh, R.; Aiyaz, M.; Divakara, S.T.; Murali, H.S.; Prakash, H.S.; Lakshmana Rao, P.V. Zearalenone induced toxicity in SHSY-5Y cells: The role of oxidative stress evidenced by N-acetyl cysteine. *Food Chem. Toxicol.* **2014**, *65*, 335–342. [CrossRef]
33. Liu, K.; Zhao, F.; Yan, J.; Xia, Z.; Jiang, D.; Ma, P. Hispidulin: A promising flavonoid with diverse anti-cancer properties. *Life Sci.* **2020**, *259*, 118395. [CrossRef] [PubMed]
34. Zhang, Z.; Zhao, Q.; Liu, T.; Zhao, H.; Wang, R.; Li, H.; Sun, H. Effect of Vicenin-2 on ovariectomy-induced osteoporosis in rats. *Biomed Pharm.* **2020**, *129*, 110474. [CrossRef]
35. Chen, X.; Yao, Z.; Peng, X.; Wu, L.; Wu, H.; Ou, Y.; Lai, J. Eupafolin alleviates cerebral ischemia/reperfusion injury in rats via blocking the TLR4/NF-κB signaling pathway. *Mol. Med. Rep.* **2020**, *22*, 5135–5144. [CrossRef] [PubMed]
36. Wu, B.; Liang, J. Pectolinarigenin promotes functional recovery and inhibits apoptosis in rats following spinal cord injuries. *Exp. Ther. Med.* **2019**, *17*, 3877–3882. [CrossRef] [PubMed]
37. Huang, L.; Huang, K.; Ning, H. Autophagy induction by hispidulin provides protection against sevoflurane-induced neuronal apoptosis in aged rats. *Biomed Pharm.* **2018**, *98*, 460–468. [CrossRef] [PubMed]
38. Pang, Q.; Zhao, Y.; Chen, X.; Zhao, K.; Zhai, Q.; Tu, F. Apigenin protects the brain against ischemia/reperfusion injury via caveolin-1/VEGF in vitro and in vivo. *Oxid. Med. Cell Longev.* **2018**, *2018*, 7017204. [CrossRef]
39. Yao, H.; Yuan, Z.; Wei, G.; Chen, C.; Duan, J.; Li, Y.; Liu, Y. Thevetiaflavone from Wikstroemia indica ameliorates PC12 cells injury induced by OGD/R via improving ROS-mediated mitochondrial dysfunction. *Mol. Med. Rep.* **2017**, *16*, 9197–9202. [CrossRef] [PubMed]
40. Bu, J.; Shi, S.; Wang, H.Q.; Niu, X.S.; Zhao, Z.F.; Wu, W.D.; Zhu, Y. Acacetin protects against cerebral ischemia-reperfusion injury via the NLRP3 signaling pathway. *Neural Regen. Res.* **2019**, *14*, 605.
41. Montesano, D.; Juan-García, A.; Mañes, J.; Juan, C. Chemoprotective effect of carotenoids from *Lycium barbarum* L. on SH-SY5Y neuroblastoma cells treated with beauvericin. *Food Chem. Toxicol.* **2020**, *141*, 111414. [CrossRef]
42. Agahi, F.; Font, G.; Juan, C.; Juan-García, A. Individual and combined effect of zearalenone derivates and beauvericin mycotoxins on SH-SY5Y Cells. *Toxins* **2020**, *12*, 212. [CrossRef]
43. Caballero-Gallardo, K.; Olivero-Verbel, J.; Corada-Fernández, C.; Lara-Martín, P.A.; Juan-García, A. Emerging contaminants and priority substances in marine sediments from Cartagena Bay and the Grand Marsh of Santa Marta (Ramsar site), Colombia. *Environ. Monit. Assess* **2021**, *193*, 1–18. [CrossRef] [PubMed]

toxins

MDPI

Article

Cytotoxicity of an Innovative Pressurised Cyclic Solid–Liquid (PCSL) Extract from *Artemisia annua*

Rosanna Culurciello [1], Andrea Bosso [1], Giovanni Di Fabio [2], Armando Zarrelli [2,3], Angela Arciello [2], Francesca Carella [1], Leonardo Leonardi [4], Laura Pazzaglia [5], Gionata De Vico [1,*] and Elio Pizzo [1,*]

1 Department of Biology, University of Naples Federico II, 80126 Naples, Italy; rosanna.culurciello@unina.it (R.C.); andrea.bosso@unina.it (A.B.); francesca.carella@unina.it (F.C.)
2 Department of Chemical Sciences, University of Naples Federico II, 80126 Naples, Italy; giovanni.difabio@unina.it (G.D.F.); armando.zarrelli@unina.it (A.Z.); anarciel@unina.it (A.A.)
3 Center for Studies on Bioinspired Agro-Environmental Technology (BAT CENTER), University of Naples Federico II, 80126 Naples, Italy
4 Department of Veterinary Medicine—Veterinary Pathology, University of Perugia, 06129 Perugia, Italy; leonardo.leonardi@unipg.it
5 Laboratory of Experimental Oncology, IRCCS Istituto Ortopedico Rizzoli, 40136 Bologna, Italy; laura.pazzaglia@ior.it
* Correspondence: gionata.devico@unina.it (G.D.V.); elipizzo@unina.it (E.P.)

Abstract: Therapeutic treatments with *Artemisia annua* have a long-established tradition in various diseases due to its antibacterial, antioxidant, antiviral, anti-malaria and anti-cancer effects. However, in relation to the latter, virtually all reports focused on toxic effects of *A. annua* extracts were obtained mostly through conventional maceration methods. In the present study, an innovative extraction procedure from *A. annua*, based on pressurised cyclic solid–liquid (PCSL) extraction, resulted in the production of a new phytocomplex with enhanced anti-cancer properties. This extraction procedure generated a pressure gradient due to compressions and following decompressions, allowing to directly perform the extraction without any maceration. The toxic effects of *A. annua* PCSL extract were tested on different cells, including three cancer cell lines. The results of this study clearly indicate that the exposure of human, murine and canine cancer cells to serial dilutions of PCSL extract resulted in higher toxicity and stronger propensity to induce apoptosis than that detected by subjecting the same cells to *Artemisia* extracts obtained through canonical extraction by maceration. Collected data suggest that PCSL extract of *A. annua* could be a promising and economic new therapeutic tool to treat human and animal tumours.

Keywords: artemisinin; anticancer effects; cytotoxicity; stress granules; alternative extraction procedures; osteosarcoma cells; HeLa cells

Key Contribution: PCSL extraction guarantees the isolation of a phytocomplex with enhanced properties.

Citation: Culurciello, R.; Bosso, A.; Di Fabio, G.; Zarrelli, A.; Arciello, A.; Carella, F.; Leonardi, L.; Pazzaglia, L.; De Vico, G.; Pizzo, E. Cytotoxicity of an Innovative Pressurised Cyclic Solid–Liquid (PCSL) Extract from *Artemisia annua*. *Toxins* **2021**, *13*, 886. https://doi.org/10.3390/toxins13120886

Received: 12 November 2021
Accepted: 8 December 2021
Published: 11 December 2021

1. Introduction

Artemisia annua is a member of the Asteraceae family, with therapeutic properties mainly related to its content in artemisinin, a sesquiterpene lactone produced in the trichomes, which presents an endoperoxide bridge indispensable for its bioactivities [1]. Artemisinin and its derivatives are active ingredients at the basis of most antimalarial treatments [2], as well as present intriguing anticancer properties related to their capability to arrest cell growth, interfere with the cell cycle and/or activate multiple cell death pathways in cancer cells [3–5]. An increasing number of evidence suggests that the anticancer properties of *A. annua* are not exclusively related to a single active component of the plant, but rather to its whole phytochemical complex, including over 600 phytochemicals dominated by sesquiterpenoids, flavonoids, coumarins, enzymes and steroids [6], which may act synergistically [7–9]. Studies also demonstrated that *A. annua* whole plant extracts could have more effective anticancer properties compared to single phytochemical compounds both in vitro and in vivo [10–12], thus supporting their promising use as new therapeutics in human and animal tumours.

However, not all the extraction methods allow to isolate from the plant fractions of active ingredients with similar properties. It is widely reported that the extraction strategies can significantly affect the potential therapeutic efficacy of obtained samples mainly due to their chemical composition [13–15]. In this context, an innovative technique, based on pressurized cyclic solid-liquid (PCSL) extraction, was recently demonstrated to be significantly more efficient than the traditional methodologies used to obtain mother tincture (European Pharmacopeia) in extracting phytochemicals from *A. annua* [2]. This technique is based on the creation of a pressure gradient, achieved through compressions and immediate decompressions of the plant tissue, which ultimately leads to the extraction of trichomes and avoiding any maceration. Thereby, this method reduces the operating time of extraction, avoids the use of solvents, and facilitates the use of crude extract, evading further manipulations and reducing toxicity, with high efficiency and low or no risks for the operator and the environment.

To date, the biological properties of *A. annua* PCSL extract have not been fully clarified; nevertheless, it is conceivable to suppose that the high levels of artemisinin and other phytochemicals detected in this product could have a positive impact on its anticancer properties compared to other preparations [2]. In this framework, the present study was mainly focused on assessment of the effects of *A. annua* PCSL extract, compared to canonical mother tincture, on four different cell lines: 1. Simian virus 40-transformed mouse cells (SVT2); 2. mouse embryonic fibroblast cells (NIH/3T3); 3. human cervical cancer cells (HeLa); 4. canine osteosarcoma cells (CRL2130). Collected results indicate that the phytocomplex achievable with this extraction strategy exerts significant cell toxicity, with a promising trend more accentuated on tumour cells, de facto opening the way to an alternative approach in the study of the plant extracts' applicability.

2. Results

2.1. Cytotoxicity Tests: PCSL Extract vs. Mother Tincture of A. annua

Cytotoxicity of serial dilutions of PCSL hydroalcoholic extract and mother tincture of *A. annua* was compared by MTT assay on normal and tumour murine fibroblasts. As shown in Figure 1, data clearly highlight a marked dose-dependent toxic effect of PCSL hydroalcoholic extract on both cell lines. Moreover, in all conditions tested (scalar doses and three different incubation times), PCSL hydroalcoholic extract ever exerts a higher toxicity if compared to mother tincture of *A. annua*, thus corroborating the initial hypothesis that the PCSL extract could present new properties and higher amounts of active ingredients.

To confirm the cytotoxic potential over time of PCSL extract, we evaluated its toxicity on further two cancer cell lines of human and canine origin (HeLa cells and canine osteosarcoma cells-CRL2130). Collected data on both cancer cell lines (see Figure 2, panels A–F) indicate a clear dose and time-dependent toxicity of PCSL hydroalcoholic extract as well as a higher toxicity if compared to that detected for similar doses from mother tincture.

2.2. PCSL Extract Induces Apoptosis and Negatively Affects the Recruitment of Stress Granules

To verify the possible PCSL extract action in the triggering of the apoptotic pathway, the differential acridine orange/ethidium bromide (AO/EB) staining was assayed on Hela and CRL2130 cells. Cells were first treated with PCSL hydroalcoholic extract (at doses equal to IC50 values determined in the MTT assays) for 48 h and then incubated with AO/EB dye mixture to detect, by fluorescence microscopy, live cells and dead apoptotic cells (see Figure 3, panels A and B). As shown in figure, both for CRL2130 (panel A) and for HeLa (panel B) treated with PCSL extracts, a significant number of apoptotic cells was detected.

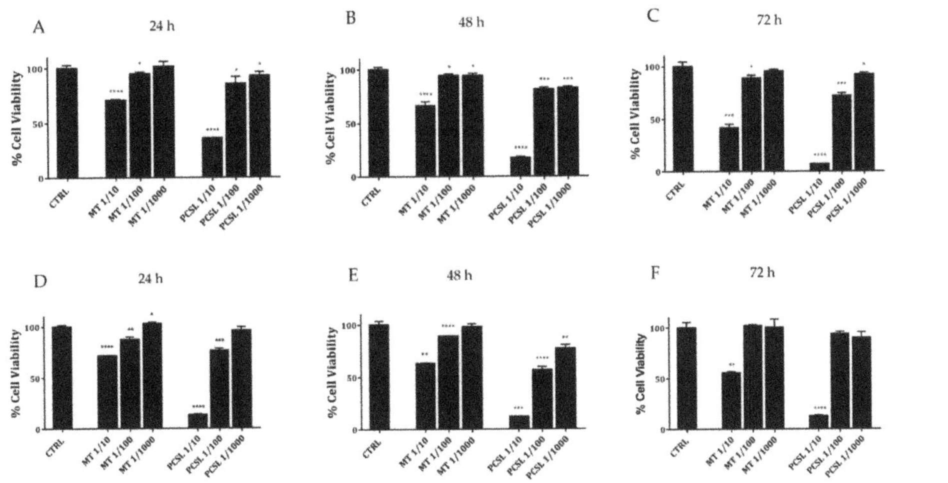

Figure 1. Cell viability of normal (3T3-panels **A–C**) and cancer (SVT2-panels **D–F**) murine fibroblasts treated with serial dilutions of PCSL hydroalcoholic extract (labelled as PCSL) or mother tincture (labelled as MT) from *A. annua* for 24, 48 and 72 h. Doses correspond to serial dilutions of the two different extracts obtained starting from similar masses of fresh flowering aerial part of *A. annua*. Experiments were performed in triplicate, and statistical analysis were carried out as described in Materials and Methods. A p value of 0.05 or less was considered statistically significant (* $p < 0.05$, ** $p < 0.01$, *** $p < 0.001$ or ****$p < 0.0001$).

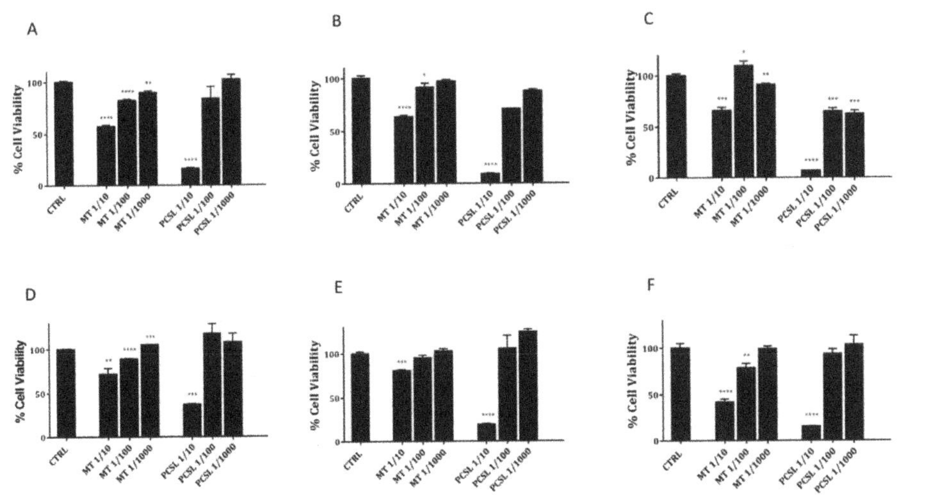

Figure 2. Cell viability of HeLa (Panels **A–C**) and CRL130 (panels **D–F**) treated with three different serial dilutions of PCSL hydroalcoholic extract or mother tincture extract (MT) for 24, 48 and 72 h. Experiments were performed in triplicate, and statistical analysis were carried out as described in materials and methods. A p value of 0.05 or less was considered statistically significant (* $p < 0.05$, ** $p < 0.01$, *** $p < 0.001$ or ****$p < 0.0001$).

A.

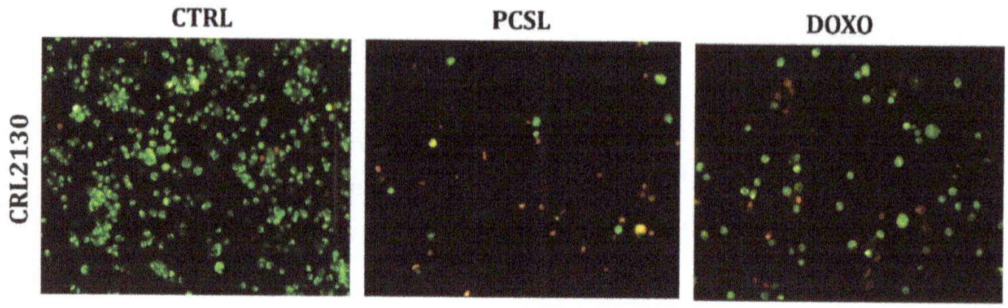

Percentage of apoptotic cells + standard deviation		
CTRL	PCSL (IC_{50} at 48h)	Doxorubicin
4.6 ± 1.46 %	79.8 ± 11.8 %	33.12 ± 9.94 %

B.

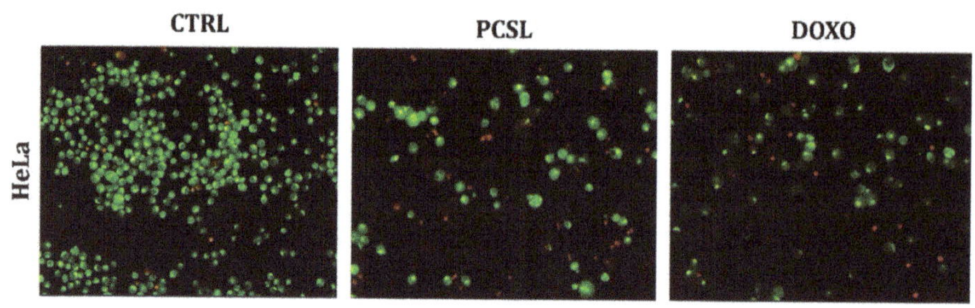

Percentage of apoptotic cells + standard deviation		
CTRL	PCSL (IC_{50} at 48h)	Doxorubicin
6.7 ± 1.08 %	39.9 ± 9.3%	25.4 ± 7.55%

Figure 3. AO/EB staining of CRL2130 (**A**) and HeLa cells (**B**) treated with *A. annua* PCSL extract (doses corresponding to IC_{50} values at 48 h). Apoptotic cells are represented as red spots, while green ones indicate viable cells. Numbers below the images correspond to the means of the percentage of apoptotic cells ± SD, considering five microscopic fields for each sample.

Based on these indications, we wanted to verify whether the action of PCSL extract could have an effect on the recruitment of stress granules, and to do this, confocal laser scanning microscopy analyses were carried out on HeLa cells pre-treated with two different dilutions of PCSL extract (see Methods). As shown in Figure 4 (panel A), the treatment with increasing doses of PCSL extract negatively affects the stress granules recruitment, both in terms of number and size, supporting the hypothesis that the toxicity of this innovative type of extract is mainly due to its ability to induce apoptosis. A further test was carried out by RT-qPCR to explore possible alterations of Bax mRNA expression, a pro-apoptotic marker. As shown in Figure 4 (panel B), a significant rise of Bax mRNA relative expression is observed at the highest concentration of PCLS extract, detected by a comparison with a housekeeping gene (GAPDH), thus suggesting the ability of this extract to affect cell viability by inducing the apoptotic pathway.

A.

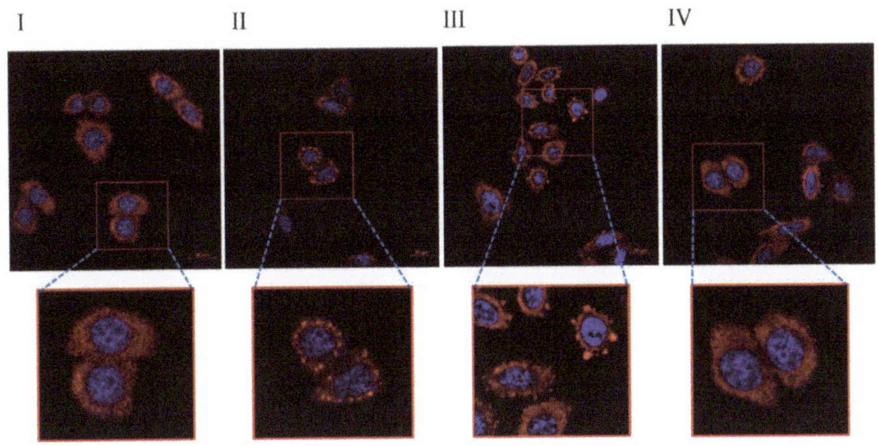

B.

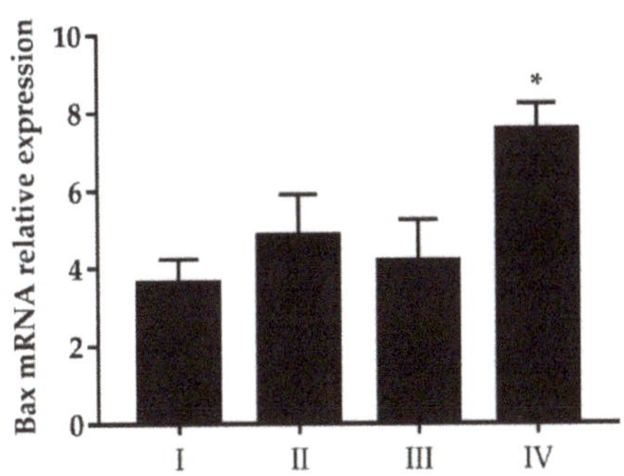

Figure 4. (**A**) Subcellular localization of PABP, a stress granule marker protein, in HeLa cells. (I) Untreated cells (negative control); (II) HeLa cells subjected to oxidative stress with 500 µM sodium arsenite (stress granules formation used as reference control); (III) HeLa cells treated with PCSL extract diluted 1:100 and (IV) HeLa cells treated with PCSL extract diluted 1:10. All treatments were carried out for 1 h at 37 °C. Fixed cells were stained with anti-human PABP (red), whereas nuclei were counterstained by using DAPI (blue). For each panel, an enlargement of the insert highlighted in red is shown. The bars indicate 20 µm. (**B**) Expression analysis by RT-qPCR of Bax gene. Bars represent relative expression levels of Bax gene normalized by GAPDH in the same experimental points shown above (I, II III and IV). A p value of 0.05 or less was considered statistically significant (* $p < 0.05$).

3. Discussion

In several reports, *A. annua* has been taken into consideration for its anti-cancer activity starting from samples obtained with canonical extraction procedures [4,16]. In the present pivotal study, we assessed the biological activity of an extract from *A. annua* obtained by an innovative technique, known as pressurized cyclic solid–liquid extraction (PCSL) [2], using two well-established in vitro models in human and animal cancer research.

There are several advantages related to the use of the Naviglio extractor as an alternative to conventional techniques: (i) the reduction of extraction times; (ii) no more than 2–24 h depending on the parts of the plant used for the extraction; (iii) the high quality of the extract, since the contact between extracted solid and liquid phases is reduced to a minimum, and therefore, the degradation of the material to be extracted is completely negligible; (iv) the obtained extract does not require filtration; (v) the negligible degradation of the extracted bioactive compounds, as the extraction process is carried out at room temperature and, if necessary, in an atmosphere of inert nitrogen; (vi) the high extraction efficiency resulting from repeated vacuum-pressure cycles [2].

The extractor applies pressure and depression cycles on the extracting liquid that is placed in contact with the solid vegetable matrix, from which the active ingredients are extracted with high efficiency, in a short time (compared to conventional maceration) and at room temperature [17]. Extracting at low temperatures is relevant in order to avoid thermal stress on different compounds. Therefore, it is possible to faithfully obtain the same composition of the substances contained in medicinal plants without inducing transformations in the bioactive constituents, which are generally the most "delicate" elements to be extracted, causing their thermolability. For this reason, the dynamic strategy of extraction, here presented, represents a promising technology useful to obtain solid matrices containing extractable compounds that are exhausted in different solvent and in their mixtures. This process is currently being used by a growing number of herbal, cosmetic, food and homeopathic companies [18,19].

Collected data highlight that the exposition of HeLa, SVT2 and CRL2130 cells to serial dilutions of extract (1:10, 1:100 and 1:1000) at three different incubation times (24, 48 and 72 h) induce strong cytotoxic effects on the cancer cells, and at the dose corresponding to dilution 1:10, this property was detectable already at short incubation times and was stable over time.

Previous studies involving HeLa cells also underlined the antineoplastic potential of *A. annua* and its active constituents (single molecules or as a whole) using canonical extraction procedures. This had an important clinical impact for the development of new therapeutic approaches in cancer treatment. Disbrow et al. [20] found that the main artemisinin derivative, dihydroartemisinin (DHA), showed a high toxicity and propensity to induce apoptosis via activation of the mitochondrial caspase pathway, in cervical cancer cells, while artemisinin had no effect [20]. Based on the results obtained on HeLa cells, Disbrow et al. tested DHA efficacy upon topical treatment of mucosal papillomavirus lesions in dogs and demonstrated a strong inhibition of viral-induced tumour formation. According to the above results, Jansen et al. [21], using an artemisinin derivative (artesunate-R), conducted a pilot study in human patients affected by cervical carcinoma. Patients treated with artesunate-R were characterized by a significant improvement of clinical manifestations and a prolonged survival time. Furthermore, observed clinical responses were associated with the gene expression downregulation of p53, EGFR, Ki-67 and CD31, which are relevant tumour proteins. Finally, a Phase I study to test the healthiness of intravaginal artesunate application in the treatment of HPV+ High Grade Cervical Intraepithelial Neoplasia is ongoing at the Sidney Kimmel Comprehensive Cancer Center at Johns Hopkins (https://clinicaltrials.gov/ct2/show/NCT02354534) (accessed on 28 September 2021) (USA).

Concerning whole extracts of *A. annua*, Efferth et al. [7] tested several of them with different phytogeographical origins and on HeLa cells. In that study, dichloromethane extract was more effective than other preparations in killing HeLa cells. Furthermore, the phytochemical composition of the extracts supported the in vitro synergistic effects of compounds that constitute the phytochemical complexes, suggesting artemisinin, arteannuin B and scopoletin as the main active components underlying the relevance of different extraction techniques on the biological effect of the extracts.

SVT2 is a malignant cell line derived from transformed murine fibroblasts and belongs to a large family of malignant cell lines transformed by SV-40, a viral agent thought to be involved in human and animal malignancy pathogenesis [22,23]. To the best of our knowledge, this is the first study on the anticancer activity of *A. annua* extract on SVT2 cells. In our study, we have compared the activity of PCSL extract on SVT2 and 3T3 cells to evaluate the selectivity of the anticancer activity of the extract. Based on obtained results, it would appear that PCSL extract from *A. annua* presents a significant dose-dependent toxicity and a detectable selectivity in an incubation time ranging from 24 to 48 h. These data reinforce the interest on PCSL extract as a promising therapeutic option, although the results should be confirmed in further research activities to optimize the dose in relation with the outcome to obtain. Among the results of our study, those obtained on CRL2130 (a canine osteosarcoma cell line) are of particular relevance. In dogs, osteosarcoma is the most common primary bone tumour that tends to occur in middle-aged specimens and in large and giant breeds, especially in the metaphyseal region of the long bones [24,25]. Canine osteosarcoma represents a relevant model in the understanding of the same disease in humans since it has genetic background, clinical symptoms, biological behaviour, treatment response and a progression comparable to that of human osteosarcoma, particularly in those occurring during adolescence [25,26]. Unfortunately, therapeutic failures are frequent in both species, with a survival time after surgery and chemotherapy only of 60% of human patients, reaching 5-year (about 20% survival time at 5 years with metastasis present at diagnosis) and two-year survival rates less than 20% in dogs (in canines with amputation and chemotherapy, the average survival time is 10–12 months) [26]. As a consequence, new therapies are desperately needed for osteosarcoma treatment to improve clinical outcome in both humans and dogs. Anti-tumour properties of artemisinin and of its derivatives have been demonstrated in both human and canine osteosarcoma cell lines [27]. Furthermore, Isani et al. [10] tested a commercial hydroalcoholic extract of *A. annua* and demonstrated that the whole phytochemical complex is more effective than artemisinin alone in killing a further canine osteosarcoma cell line (D-17). In our study, the antineoplastic activity detected for PCSL extract on canine osteosarcoma cell lines confirm that *A. annua* extract has a therapeutic potential for these tumours. In our case, a generic activation of apoptosis pathways was demonstrated in vitro after exposing cells to PCSL extract as suggested by results obtained by the differential staining assay of normal and apoptotic cells using a dye mixture with acridine orange/ethidium bromide (AO/EB) and was further supported by data of stress granules and RT-qPCR to explore possible alterations of pro-apoptotic Bax mRNA expression [28].

Stress granules are transient cytoplasmic aggregates, based on RNAs and proteins, that allow a rapid retrieval of cellular homeostasis, following stress stimuli [29]. Several reports indicate an intimate link between stress granules and cancer cells and, in particular, the ability of these complexes to inhibit trigger of apoptosis, thus integrating oncogenic signalling with the increase of cancer cell fitness [30,31]. Data of the present study show that PCSL extract negatively affects the formation of the stress granules, both in terms of number and size, strongly supporting the hypothesis that the toxicity of this innovative type of extract is mainly due to its ability to induce apoptosis. Some secondary metabolites from *A. annua*, such as scopoletin, might be responsible for the activation of a classic mitochondrial apoptosis pathway in cancer cells [32,33].

However, canine and human osteosarcomas, in line with cancer cells hallmarks, are also known to overexpress membrane transferrin receptors, which make cells particularly susceptible to a particular form of iron-dependent non-apoptotic cell death, known as Ferroptosis, the latter of which is activated by artemisinin [34]. It is, therefore, possible that the phytochemicals present in the PCSL extract of *A. annua* activate multiple molecular cell death pathways in cancer cells. Hence, further studies are necessary to clarify the nature and the number of death pathways involved in the antineoplastic activity of the extract used in our study. In conclusion, in the current focus on discovering innovative therapeutics, our preliminary results suggest that PCSL extract of *A. annua* could be a promising and economic new tool to treat animal tumours refractory to conventional agents.

4. Materials and Methods

4.1. Plant Material

Summer *A. annua* plants were harvested near Caserta (Southern Italy), characterized and deposited in the herbarium of University of Naples Federico II (HERBAZLS 280815).

4.2. Chemicals and Materials

Artemisinin (CAS: 63968–64–9; purity > 98%), scopoletin (CAS: 92–61–5; purity > 98%), HPLC-grade acetonitrile, methanol and phosphoric acid were purchased from Sigma Aldrich (Milan, Italy). Ultrapure quality water, generated in the laboratory using a Milli-Q water purification system (Millipore, Bedford, MA, USA), was used throughout the experiments.

4.3. Pressurized Cyclic Solid–Liquid (PCSL) Extractor

The Naviglio Extractor, a name that follows that of the manufacturer and the relative principle, allows to extract the compounds that are not chemically bound to a vegetable matrix with a suitable solvent, exerting a negative pressure gradient between the outside and the inside until the initial equilibrium conditions are quickly reached [17]. The extractor applies pressure and depression cycles on the extracting liquid that is placed in contact with the solid vegetable matrix, from which the active ingredients are extracted with high efficiency, in a short time (compared to conventional maceration) and at room temperature [35]. The extraction therefore takes place in conditions that do not damage the thermolabile substances, which are generally the most delicate and subject to transformation/degradation due to the effect of temperature, faithfully preserving the composition of an extract of natural origin in relation to all its active ingredients. The dynamic solid–liquid rapid extractor shows an innovative solid–liquid extraction technology that allows the solid matrices containing extractable substances to be exhausted in pure or mixed solvents with different polarity. This process is currently being used by a growing number of cosmetic, food, homeopathic and herbal companies.

4.4. Hydroalcoholic Extracts by PCSL Extractor

The PCSL extraction method was carried out using a Naviglio Extractor equipped with a housing of the vegetable matrix all in steel. It used ethanol:H_2O (3:2, *v/v*, 550 mL) or H_2O (540 mL) as extraction solvent; 45 and 35 g of fresh plant, respectively, and a pressure of about 10 bar and an immediate decompression at about 0–1 bar of pressure. An extraction cycle comprises the static and dynamic phases of 3 and 1 min, respectively. The extraction lasted 3 and 24 h in hydroalcoholic solution, respectively.

4.5. Qualitative-Quantitative Analysis of Artemisinin and Scopoletin by HPLC

Analysis of artemisinin and scopoletin was performed using a Shimadzu LC-20 HPLC chromatograph (see Figure 5), equipped with a spectrofluorometric detector Shimadzu RF10AXL (Milan, Italy) and an analytical column RP18 (particle size 5 µm, 150 × 4.6 mm i.d., Waters, Milford, MA, USA). The amount of artemisinin and scopoletin was determined under isocratic conditions using a mixture of methanol:H_2O:H_3PO_4 (50:50:0.1, *v/v/v*) and a flow of 1.5 mL/min. The excitation and emission wavelengths for artemisinin were 210 and 246 nm, respectively, while the excitation and emission wavelengths for scopoletin were 430 and 460 nm, respectively. One mL of the hydroalcoholic extract was diluted 20 times with milli-Q water, and then 100 µL of each solution was mixed with an equal volume of the mobile phase and analysed by HPLC in triplicate.

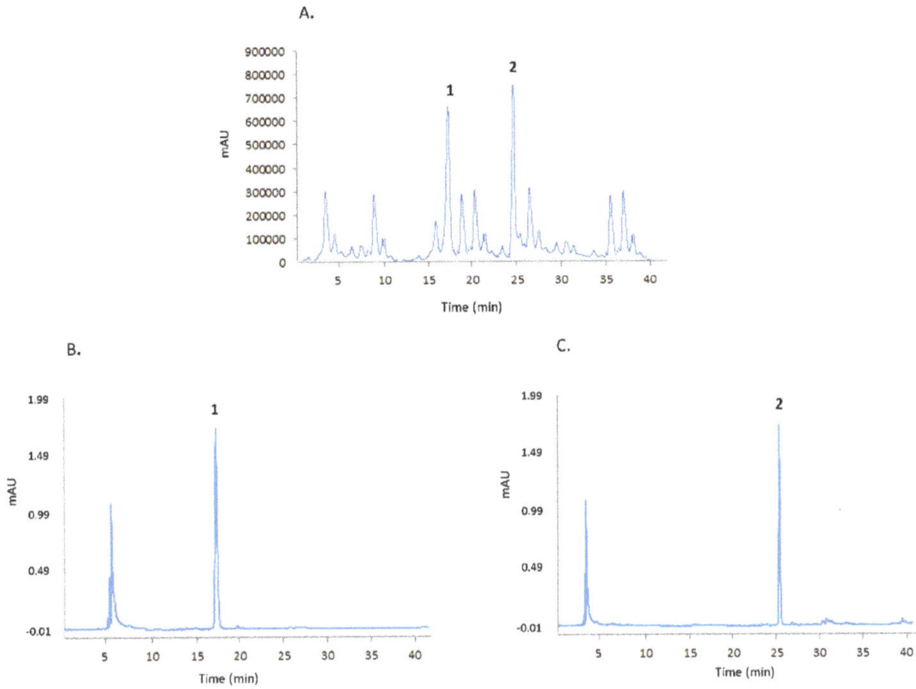

Figure 5. (**A**) Hydroalcoholic extract by PCSL extractor; (**B**) scopoletin (**1**); (**C**) artemisinin (**2**).

4.6. Preparation of Stock and Working Solutions

For the determination of scopoletin, a calibration curve was constructed using eight different solutions containing from 3 to 50 mg of standard, suitably diluting equal aliquots of a stock solution at a concentration of 1.0 mg/mL. The solvent used for the stock solution and the subsequent dilutions was ethanol, and care was taken to keep the fractions in the dark and at 4 °C before use. The prepared solutions were analysed spectrofluorimetrically against the solvent blank (absolute ethanol). For each solution, from the measurement of the fluorescence intensity (X) as a function of the quantity of standard present (Y), it was possible to obtain a calibration curve of equation equal to $Y = 7.36 \times 10^{-5}$ X, with a value of the coefficient correlation (R) equal to 0.999. The same procedure was used to obtain the calibration curve for artemisinin, starting from a stock solution with a concentration of the commercial standard equal to 1.0 mg/mL and diluted solutions with an amount of standard equal to 0.1–10 μg/mL. The line equation for the standard curve was $Y = 6.11 \times 10^{-4}$ X and correlation coefficient (R) value was 0.999.

4.7. Cell lines, Cell Viability Experiments and Apoptosis Assays

Simian virus 40-transformed mouse cells (SVT2), human cervical cancer cells (HeLa) and canine osteosarcoma cells (CRL2130) were used to test the anticancer activity of PCSL extract of A. annua. Mouse embryonic fibroblast cells (NIH/3T3) were used as non-tumour control cells.

SVT2 (Balb/c 3T3 mouse cells transformed by simian virus 40), NIH/3T3 (Balb/c 3T3 mouse embryonic fibroblasts), HeLa (human cervical carcinoma cells) were supplied by ATCC (www.atcc.org) (accessed on 17 November 2021). DAN CRL2130 (canine osteogenic sarcoma cells) were kindly provided by Dr. Laura Pazzaglia from Istituto Ortopedico Rizzoli (Bologna, Italy). SVT2, 3T3 and HeLa cells were maintained in DMEM supplemented with 10% FBS, 1% pen/strep and 1% L-Glu, while CRL2130 cells were cultured in EMEM supplemented with 10% FBS, 1% pen/strep and 1% L-Glu and 0.4 mg/mL G418.

To evaluate the cytotoxic potential of the PCSL extract from A. annua, 5×10^3 cells were seeded in 96-well plates, and after 24 h of growth, increasing amounts of extract were added. Cells viability was assessed at three different incubation times (24, 48 and 72 h) by MTT method. The ratio between absorbance values of treated cells and those concerning untreated cells (positive control) indicates the rate of cell proliferation (in our case expressed as cell viability percentage). Apoptosis was detected by differential acridine orange/ethidium bromide (AO/EB) staining [36] on HeLa and CRL2130 cells by using a procedure described in [37].

4.8. Immunofluorescence Experiments

The effect of *A. annua* PCSL extract on stress granules recruitment was tested in HeLa cells. Briefly, 2×10^4 HeLa cells were seeded on coverslips, cultured for 24 h and then treated at 37 °C for 1 h with two different dilutions of extract (1:10 and 1:100). The treatment of HeLa cells with 500 μM of sodium arsenite for 1 h at 37 °C was considered as a positive control of stress granules recruitment.

Cells were finally fixed in 4% paraformaldehyde for 15 min, permeabilized in 0.1% Triton-X100 for 5 min and blocked with 1% BSA (bovine serum albumin) for 20 min. Subsequently, cells were incubated with mouse anti-PABP (P6246-Sigma-Aldrich), diluted 1:750 in BSA 1% for 1 h, and then with the secondary antibody goat anti-Mouse (DyLight ® 594 Conjugate). Nuclear staining of the cells was obtained by incubation with DAPI (Molecular Probes, Invitrogen, Italy) diluted 1:2000 in PBS for 5 min at room temperature. After washing, coverslips were mounted in Mowiol ® 4–88 (Sigma Aldrich, Milan, Italy) on microscope slides. Confocal microscopy observations were performed using Zeiss Confocal Microscope LSM 900 and a 63× oil objective.

4.9. Real-Time qPCR Analysis

Expression of Bax gene, a pro-apoptotic marker, was evaluated by Real-Time qPCR in HeLa cells subjected to treatment with two alternative doses of PCSL extract. In brief, 3×10^5 cells were plated, and after 24 h, exposed to PCLS extract (1:10 or 1:100) for 1 h at 37 °C. Total RNA was then extracted using TRIzol ™ Reagent (Invitrogen ™), according to the manufacturer's instructions, quantified by Nanodrop 8000 (Thermo-Scientific) and finally retro-transcripted to cDNA with SuperScript ™ IV VILO ™ Master Mix (Invitrogen ™). RT-PCRq was performed by StepOnePlus ™ Real-Time PCR System (Applied Biosystems ™) using the primers listed in Table 1. Expression of Bax gene was normalized with respect to the housekeeping gene encoding GAPDH.

Table 1. Primer sequences of selected genes.

Scheme	Forward	Reverse
Bax [38]	5′-TGCTTCAGGGTTTCATCCAG-3′	5′ -GGCGGCAATCATCCTCTG-3′
GAPDH	5′-CACCACACTGAATCTCCCCT-3′	5′-TGGTTGAGCACAGGGTACTT- 3′

4.10. Statistical Analysis

Data were analysed with GraphPad Prism software (version 5.0-GraphPad Inc., San Diego, CA, USA) by using Student's *t* test. A *p* value of 0.05 or less was considered statistically significant (* $p < 0.05$, ** $p < 0.01$, *** $p < 0.001$ or ****$p < 0.0001$).

Author Contributions: Conceptualization, G.D.V. and E.P.; Data curation, G.D.V. and E.P.; Formal analysis, R.C., A.B., G.D.F., A.A. and F.C.; Investigation, R.C., A.B., G.D.F., A.A. and F.C.; Methodology, G.D.V. and E.P.; Project administration, G.D.V. and E.P.; Resources, A.Z., L.P., G.D.V. and E.P.; Supervision, L.L., G.D.V. and E.P.; Validation, A.Z., G.D.V. and E.P.; Writing—original draft, A.Z., G.D.V. and E.P.; Writing—review & editing, A.Z., G.D.V. and E.P. All authors have read and agreed to the published version of the manuscript.

Funding: This research was performed thanks to the economic efforts drawn from the depart-mental research funds of all authors.

Institutional Review Board Statement: Not applicable.

Informed Consent Statement: Not applicable.

Data Availability Statement: Not applicable.

Acknowledgments: This study was supported by AIPRAS-Onlus (Associazione Italiana per la Promozione delle Ricerche sull'Ambiente e la Salute umana).

Conflicts of Interest: The authors declare no conflict of interest.

References

1. Maes, L.; Van Nieuwerburgh, F.C.W.; Zhang, Y.; Reed, D.W.; Pollier, J.; Vande Casteele, S.R.F.; Inzé, D.; Covello, P.S.; Deforce, D.L.D.; Goossens, A. Dissection of the phytohormonal regulation of trichome formation and biosynthesis of the antimalarial compound artemisinin in *Artemisia annua* plants. *New Phytol.* **2011**, *189*, 176–189. [CrossRef]
2. Zarrelli, A.; Pollio, A.; Aceto, S.; Romanucci, V.; Carella, F.; Stefani, P.; De Natale, A.; De Vico, G. Optimisation of artemisinin and scopoletin extraction from *Artemisia annua* with a new modern pressurised cyclic solid–liquid (PCSL) extraction technique. *Phytochem. Anal.* **2019**, *30*, 564–571. [CrossRef] [PubMed]
3. Konstat-Korzenny, E.; Ascencio-Aragón, J.; Niezen-Lugo, S.; Vázquez-López, R. Artemisinin and Its Synthetic Derivatives as a Possible Therapy for Cancer. *Med. Sci.* **2018**, *6*, 19. [CrossRef]
4. Slezakova, S.; Ruda-Kucerova, J. Anticancer activity of artemisinin and its derivatives. *Anticancer Res.* **2017**, *37*, 5995–6003. [CrossRef]
5. Wong, Y.K.; Xu, C.; Kalesh, K.A.; He, Y.; Lin, Q.; Wong, W.S.F.; Shen, H.M.; Wang, J. Artemisinin as an anticancer drug: Recent advances in target profiling and mechanisms of action. *Med. Res. Rev.* **2017**, *37*, 1492–1517. [CrossRef]
6. Brown, G.D. The biosynthesis of artemisinin (Qinghaosu) and the phytochemistry of *Artemisia annua* L. (Qinghao). *Molecules* **2010**, *15*, 7603–7698. [CrossRef] [PubMed]
7. Efferth, T.; Koch, E. Complex Interactions between Phytochemicals. The Multi-Target Therapeutic Concept of Phytotherapy. *Curr. Drug Targets* **2010**, *12*, 122–132. [CrossRef] [PubMed]
8. Weathers, P.J.; Reed, K.; Hassanali, A.; Lutgen, P.; Engeu, P.O. *Whole Plant Approaches to Therapeutic Use of* Artemisia annua *L. (Asteraceae)*; Springer: Heidelberg, Germany, 2014.
9. Singh, N.P.; Ferreira, J.F.S.; Park, J.S.; Lai, H.C. Cytotoxicity of ethanolic extracts of *Artemisia annua* to molt-4 human leukemia cells. *Planta Med.* **2011**, *77*, 1788–1793. [CrossRef]
10. Isani, G.; Bertocchi, M.; Andreani, G.; Farruggia, G.; Cappadone, C.; Salaroli, R.; Forni, M.; Bernardini, C. Cytotoxic Effects of *Artemisia annua* L. And pure artemisinin on the D-17 canine osteosarcoma cell line. *Oxid. Med. Cell. Longev.* **2019**, *2019*, 1615758. [CrossRef]
11. Bilia, A.R.; Melillo de Malgalhaes, P.; Bergonzi, M.C.; Vincieri, F.F. Simultaneous analysis of artemisinin and flavonoids of several extracts of *Artemisia annua* L. obtained from a commercial sample and a selected cultivar. *Phytomedicine* **2006**, *13*, 487–493. [CrossRef]
12. Breuer, E.; Efferth, T. Treatment of Iron-Loaded Veterinary Sarcoma by *Artemisia annua*. *Nat. Products Bioprospect.* **2014**, *4*, 113–118. [CrossRef]
13. Iqbal, S.; Younas, U.; Chan, K.W.; Zia-Ul-Haq, M.; Ismail, M. Chemical composition of *Artemisia annua* L. leaves and antioxidant potential of extracts as a function of extraction solvents. *Molecules* **2012**, *17*, 6020–6032. [CrossRef]
14. Vidic, D.; Čopra-Janićijević, A.; Miloš, M.; Maksimović, M. Effects of Different Methods of Isolation on Volatile Composition of *Artemisia annua* L. *Int. J. Anal. Chem.* **2018**, *2018*, 9604183. [CrossRef] [PubMed]
15. Nahar, L.; Guo, M.; Sarker, S.D. A review on the latest advances in extraction and analysis of artemisinin. *Phytochem. Anal.* **2020**, *31*, 5–14. [CrossRef]
16. Lang, S.J.; Schmiech, M.; Hafner, S.; Paetz, C.; Steinborn, C.; Huber, R.; El Gaafary, M.; Werner, K.; Schmidt, C.Q.; Syrovets, T.; et al. Antitumor activity of an *Artemisia annua* herbal preparation and identification of active ingredients. *Phytomedicine* **2019**, *62*, 152962. [CrossRef]
17. Naviglio, D. Naviglio's principle and presentation of an innovative solid-liquid extraction technology: Extractor Naviglio®. *Anal. Lett.* **2003**, *36*, 1647–1659. [CrossRef]
18. Caprioli, G.; Iannarelli, R.; Sagratini, G.; Vittori, S.; Zorzetto, C.; Sánchez-Mateo, C.C.; Rabanal, R.M.; Quassinti, L.; Bramucci, M.; Vitali, L.A.; et al. Phenolic acids, antioxidant and antiproliferative activities of Naviglio®extracts from *Schizogyne sericea* (Asteraceae). *Nat. Prod. Res.* **2017**, *31*, 515–522. [CrossRef]
19. De Marco, A.; Luongo, G.; Di Marino, C.; De Tommaso, G.; Di Fabio, G.; Zarrelli, A. Silymarin from Silybum marianum by Naviglio's extractor: A new and very efficient approach. *Nat. Prod. Res.* **2021**, *35*, 2621–2627. [CrossRef]

20. Disbrow, G.L.; Baege, A.C.; Kierpiec, K.A.; Yuan, H.; Centeno, J.A.; Thibodeaux, C.A.; Hartmann, D.; Schlegel, R. Dihydroartemisinin is cytotoxic to papillomavirus-expressing epithelial cells in vitro and in vivo. *Cancer Res.* **2005**, *65*, 10854–10861. [CrossRef] [PubMed]
21. Jansen, F.H.; Adoubi, I.; J.C., C.K.; De Cnodder, T.; Jansen, N.; Tschulakow, A.; Efferth, T. First study of oral artenimol-R in advanced cervical cancer: Clinical benefit, tolerability and tumor markers. *Anticancer Res.* **2011**, *31*, 4417–4422. [PubMed]
22. Pipas, J.M. SV40: Cell transformation and tumorigenesis. *Virology* **2009**, *384*, 294–303. [CrossRef] [PubMed]
23. Rotondo, J.C.; Mazzoni, E.; Bononi, I.; Tognon, M.; Martini, F. Association Between Simian Virus 40 and Human Tumors. *Front. Oncol.* **2019**, *9*, 670. [CrossRef]
24. Mueller, F.; Fuchs, B.; Kaser-Hotz, B. Comparative biology of human and canine osteosarcoma. *Anticancer Res.* **2007**, *27*, 155–164. [PubMed]
25. Rankin, E.B.; Wu, C.; Khatri, R.; Wilson, T.L.S.; Andersen, R.; Araldi, E.; Rankin, A.L.; Yuan, J.; Kuo, C.J.; Schipani, E.; et al. The HIF signaling pathway in osteoblasts directly modulates erythropoiesis through the production of EPO. *Cell* **2012**, *149*, 63–74. [CrossRef] [PubMed]
26. Wilson-Robles, H.; Franks, K.; Pool, R.; Miller, T. Characterization of five newly derived canine osteosarcoma cell lines. *BMC Vet. Res.* **2019**, *15*, 357. [CrossRef]
27. Hosoya, K.; Murahari, S.; Laio, A.; London, C.A.; Couto, C.G.; Kisseberth, W.C. Biological activity of dihydroartemisinin in canine osteosarcoma cell lines. *Am. J. Vet. Res.* **2008**, *69*, 519–526. [CrossRef]
28. Pawlowski, J.; Kraft, A.S. Bax-induced apoptotic cell death. *Proc. Natl. Acad. Sci. USA* **2000**, *97*, 529–531. [CrossRef]
29. Buchan, J.R.; Parker, R. Eukaryotic Stress Granules: The Ins and Out of Translation What are Stress Granules? *Mol. Cell* **2009**, *36*, 932. [CrossRef]
30. Takahashi, M.; Higuchi, M.; Matsuki, H.; Yoshita, M.; Ohsawa, T.; Oie, M.; Fujii, M. Stress Granules Inhibit Apoptosis by Reducing Reactive Oxygen Species Production. *Mol. Cell. Biol.* **2013**, *33*, 815–829. [CrossRef] [PubMed]
31. Gao, X.; Jiang, L.; Gong, Y.; Chen, X.; Ying, M.; Zhu, H.; He, Q.; Yang, B.; Cao, J. Stress granule: A promising target for cancer treatment. *Br. J. Pharmacol.* **2019**, *176*, 4421–4433. [CrossRef]
32. Tian, Q.; Wang, L.; Sun, X.; Zeng, F.; Pan, Q.; Xue, M. Scopoletin exerts anticancer effects on human cervical cancer cell lines by triggering apoptosis, cell cycle arrest, inhibition of cell invasion and PI3K/AKT signalling pathway. *J. BUON* **2019**, *24*, 997–1002.
33. Shi, Z.; Chen, L.; Sun, J. Novel Scopoletin Derivatives Kill Cancer Cells by Inducing Mitochondrial Depolarization and Apoptosis. *Anticancer Agents Med. Chem.* **2021**, *21*, 1774–1782. [CrossRef] [PubMed]
34. De Vico, G.; Martano, M.; Maiolino, P.; Carella, F.; Leonardi, L. Expression of transferrin receptor-1 (TFR-1) in canine osteosarcomas. *Vet. Med. Sci.* **2020**, *6*, 272–276. [CrossRef] [PubMed]
35. Naviglio, D.; Scarano, P.; Ciaravolo, M.; Gallo, M. Rapid solid-liquid dynamic extraction (RSLDE): A powerful and greener alternative to the latest solid-liquid extraction techniques. *Foods* **2019**, *8*, 245. [CrossRef]
36. Kasibhatla, S. Acridine Orange/Ethidium Bromide (AO/EB) Staining to Detect Apoptosis. *Cold Spring Harb. Protoc.* **2006**, *2006*, 4493. [CrossRef]
37. Landi, N.; Ragucci, S.; Culurciello, R.; Russo, R.; Valletta, M.; Pedone, P.V.; Pizzo, E.; Di Maro, A. Ribotoxin-like proteins from Boletus edulis: Structural properties, cytotoxicity and in vitro digestibility. *Food Chem.* **2021**, *359*, 129931. [CrossRef] [PubMed]
38. Siddiqui, M.A.; Wahab, R.; Ahmad, J.; Farshori, N.N.; Saquib, Q.; Khan, S.T.; Al-Salem, A.M.; Musarrat, J.; Al-Khedhairy, A.A. Zinc oxide nanoparticles: Mechanism(s) of cell death induced in human epidermoid larynx cell line (HEp-2). *Nanosci. Nanotechnol. Lett.* **2017**, *9*, 573–582. [CrossRef]

MDPI
St. Alban-Anlage 66
4052 Basel
Switzerland
Tel. +41 61 683 77 34
Fax +41 61 302 89 18
www.mdpi.com

Toxins Editorial Office
E-mail: toxins@mdpi.com
www.mdpi.com/journal/toxins

www.ingramcontent.com/pod-product-compliance
Lightning Source LLC
Chambersburg PA
CBHW062100220926
51850CB00010B/2837